A WAY WITH
NUMBERS

A WAY WITH NUMBERS

A practical start to improve numeracy skills

TERRY RILEY

<u>To my Father - T.R.</u>

Published by BBC Books,
a division of BBC Enterprises Limited,
Woodlands, 80 Wood Lane, London W12 0TT
First published 1990
© Terry Riley 1990
ISBN 0 563 36163 8
Designed by Peartree Design Associates
Illustrations by Will Giles and Sandra Pond
Set in ITC Century Book and Frutiger by
Goodfellow and Egan Ltd, Cambridge
Printed and bound in Great Britain by
Ebenezer Baylis Ltd, Worcester
Cover printed by Richard Clay Ltd, St Ives Plc.

The author

Terry Riley has worked in Basic Skills for 20 years as a teacher, advisor and administrator, working both locally and throughout the United Kingdom. He is now a freelance Training Consultant and has also published other books on numeracy. Terry Riley is series consultant to the TV series *A Way With Numbers* and *Sum Chance*.

This book accompanies the BBC TV series *A Way With Numbers* (produced by George Auckland), first broadcast on BBC 1 starting in October 1990.

Published to accompany a series of programmes prepared in consultation with the Continuing Education Advisory Council.

Contents

Acknowledgements

This book would not have been possible without the collaboration of Martin Lang, an experienced adult numeracy tutor who has contributed ideas and examples throughout the book. His contribution has been invaluable and I am particularly indebted to him for much of the material which forms Section 20.

I would also like to thank George Auckland, senior producer of the two TV numeracy series, for his belief that a book to accompany the *A Way With Numbers* programmes was worthwhile. Thanks also to those tutors and students who, often unbeknowns, helped shape the content of this little book.

Needless to say any errors and shortcomings are very much my own.

Dear Reader,

This book is an introduction to working with numbers. It accompanies the TV series of the same title and each of the 20 sections in it reflects the contents of the 20 individual TV programmes. However, it can also be used on its own or as part of a learning programme.

You may wish to go through the book section by section but it can also be used as a reference book. Each of the sections is split into parts:

LOOKING AT NUMBER looks at how we can handle numbers and explains how we can use numbers to solve problems.

TIP are short cuts we can use to help speed up calculations or 'sums', but remember it is better to know why these work than just remember the tip.

Now try these are examples of problems for you to try out yourself. The answers are at the back of the book on pages 92-95. If you get an answer that isn't 'right' don't despair! It's very easy to make a mistake. Check back my answer to the Looking at Number explanation to see why you think I've come up with my answer. We can learn a lot by checking back. If you still get a different answer why not talk it through with someone else? (If you check it with a few people and you all get the same answer which is different to mine, the chances are that I've made the mistake!)

SIDENUMBER are at the end of each section. There's more to numbers than sums! Sidenumber give some background info. to the fascinating world of numbers.

I hope you find this book useful. I also hope that you see that, though there's more to number than meets the eye, there is 'A Way With Numbers.'

Best Wishes
TERRY RILEY

As Long as a Piece of String

Numbers are important in all our lives. We use them to know which bus to catch, what time a television programme is on and how much something costs. Everywhere we look we see numbers. It is becoming increasingly important to understand numbers and how to use them.

In this section we will look at how we use numbers for measuring. We don't always have to use a ruler or tape measure. Often we don't need to give an exact measurement, just an estimate. If someone asked 'How far is it to the cinema?' we would just give an estimate based on our experience. Even when we deal with exact measurements such as where to saw a piece of wood to make a shelf, it is often better to mark the length directly on to the wood.

But there are times when we need to be able to read and record measurements. To do this we need to use something, such as a ruler or tape measure, which has a scale marked on it. In the UK we use two different systems for measuring lengths. Nowadays many things are measured using the metric system, but the older imperial system is still commonly used.

For many purposes it does not matter which system you use, the important thing is not to confuse them.

LOOKING AT NUMBER

To be confident when using numbers you need to understand why we count and write numbers as we do. We use numbers to pass on and receive information. If you wanted to buy an egg for everyone in the family you could take them all to the shop and ask for an egg for each of them. It is obviously not very sensible. One way to

avoid this is to put a mark on a piece of paper for each member of the family and take that to the shop.

This method would be perfectly all right for small numbers – but what if we were shopping for our neighbours as well?

It is very easy to make a mistake. To make the marks easier to understand we could put them in groups.

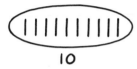

This is what we do when we count. We group things in tens.

The words we use for numbers show this. Sixty means six groups of ten, seventy means seven groups of ten. So sixty-five means six groups of ten and five extra. Ten groups of ten make a hundred, and ten groups of a hundred are a thousand.

Since it would not be very convenient to have to make a separate mark for each item we are counting we use special symbols, which are usually called digits.

1 2 3 4 5 6 7 8 9 0

So instead of putting three marks we write the digit 3.

We only have these ten digits to write any number, however large. To do this we make use of the way that we count in groups. To write down seventy-three (seven groups of ten and three extra) we would write 73, which is the number of groups of ten (7) followed by the number of extra ones (3). For even bigger numbers we write the number of hundreds followed by the number of tens and then any extra ones. So 243 means two groups of a hundred, four groups of ten, and then three extra. With this method of writing numbers the place of the digit in the

number shows its value, so it is called place value.

The 3 in	**23**	**means three**
but the 3 in	**31**	**means thirty**
and the 3 in	**327**	**means three hundred**

The digits at the left-hand side of a number have more value than those to the right.

With place value it is important to make sure that a digit is in its correct place in a number. To do this we sometimes have to use a 0 to show that there is nothing to go in a particular place. 402 means four hundreds, no tens, and two extra. If we just wrote 42, that would mean four tens and two extra; a completely different number!

Now try these
Write these numbers using digits. The answers are on page 92.

1 (a) **Twenty-five**
   (b) **Forty**
   (c) **Three hundred and sixty-two**
   (d) **Two hundred and fifty**
   (e) **Five hundred and six**
   (f) **Nine hundred**

Write these numbers using words. The answers are on page 92.

2 (a) **31**   (b) **60**   (c) **295**
   (d) **340**   (e) **409**   (f) **700**

LOOKING AT NUMBER
We often need to add numbers. When we are adding numbers what we are doing is using a quick way of putting two or more groups of things together and counting them. We would normally do this by using numbers rather than the actual items. If

we had two pints of milk one day and three the next, we could line the bottles up and count them to find how many we had altogether. Another way would be to make two marks for the first day and three for the next and count the marks. The easiest way is to remember that two and three make five altogether. It doesn't matter what you are talking about – bottles of milk, pound coins or elephants – if you have got two and three of them, you have five altogether. To add quickly and accurately it helps if you can remember how many you have when you add any two digits together. You may well already know this, but if you don't it is useful to write them down in order like this.

1 + 0 = 1   (We use the symbol + to mean add, and the symbol
1 + 1 = 2   = to show the answer.)
1 + 2 = 3

and so on to

9 + 9 = 18

It doesn't matter in what order you add two numbers, you still get the same answer: 2 + 4 = 6 and 4 + 2 = 6.

When we are adding large numbers it would be very long-winded if we had to count the numbers up each time, and it would be impossible to remember what every possible pair of numbers added up to. Because of the way we write numbers adding up large numbers isn't really very difficult. To add 26 and 43 we write it like this:

tens	ones
2	6
+ 4	3
6	9

Here we have written the numbers one below the other making sure that the tens places line up and the ones places line up. We then need to see how many ones we have altogether (6 + 3 = 9) and how many tens (2 + 4 = 6). In total we have six tens and nine ones, making 69.

The main problem with adding occurs when one of the columns adds up to more than 9. What happens if we want to add 27 and 45?

tens	ones
2	7
+ 4	5

We begin with the right-hand column.

$$7 + 5 = 12$$

12 is one ten and 2 ones. So we write 2 ones down in the ones column and carry the one ten into the tens column where we add it on to the numbers already in the tens column.

$$27 + 45 = 72$$

Now try these
Add up these numbers. Remember to keep them in columns of ones and tens. The answers are on page 92.

3   (a) 6   (b) 24   (c) 32
     + 8   + 7   + 45

    (d) 56   (e) 19   (f) 8
     + 9   + 62   + 76

LOOKING AT NUMBER –
How We Measure
We measure lengths by using a scale such as a ruler which is marked off into equal-length sections. We put this scale alongside what we are measuring and see how many sections make up its length.

Rulers and tape measures are marked with either metric or imperial measurements, and often with both.

In the metric system all measurements are based on tens and hundreds and thousands, so they fit in with our number system.

The main units in the metric system are

the metre (often written as m.) and the centimetre (cm.).

A centimetre is this long: ▬

A hundred centimetres make up one metre.

If you measure all the way round the edge of this page you will find that it is about 1 metre. To see how long this is, spread out a couple of sheets of newspaper edge to edge and then mark off the width of this page twice and the length of this page twice and that is about 1 metre altogether.

The main units in the imperial system are the inch (in.), the foot (ft.) and the yard (yd.).

An inch is this long: ▬▬

Twelve inches make up one foot. A foot is a bit more than the length of this page.

Three feet make one yard. A yard is a bit less than the metre that you marked on the newspaper.

LOOKING AT NUMBER

We don't often have to convert a measurement from one system to another – metres and centimetres to feet and inches or the other way round. The most important thing is to have a 'feel' for one system of measurement and be confident when using it. Try the other system only when you are completely happy with the one you are used to.

Now try these

Measure the lengths of these lines: the first three in metric, and the other three in imperial. The answers are on page 92.

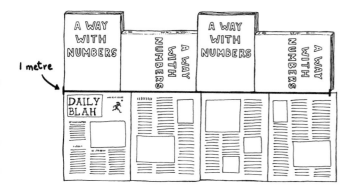

4

LOOKING AT NUMBER

Although metres and centimetres are the most commonly used metric units, there are others. For measuring long distances there is the kilometre (km.) which is 1000 metres. It takes about 10 minutes to walk one kilometre at a normal walking speed.

For measuring small amounts, or for measuring more accurately there is the millimetre (mm.). There are 10 millimetres in a centimetre. A millimetre is the distance between these lines.

In the imperial system we use miles for measuring long distances.

It is important, whichever system you are using, to measure in a suitable unit for the job. To measure the distance between London and Edinburgh it would be silly to measure it in centimetres or inches; you would use kilometres or miles.

TIP

Some rulers and measures have a piece at the end which is not part of the scale: the scale actually starts at the first mark. Other measures, such as tape measures and joiners' rules, normally start at the very end, and that is where you start measuring from. On rulers and tape measures, the mark you start measuring from is the 0 mark but normally the number 0 is not written alongside this mark.

TIP

To add up several numbers it is quicker to group the figures into tens and count them. It is worth remembering this number table:

$$1 + 9 = 10$$
$$2 + 8 = 10$$
$$3 + 7 = 10$$
$$4 + 6 = 10$$
$$5 + 5 = 10$$
$$6 + 4 = 10$$
$$7 + 3 = 10$$
$$8 + 2 = 10$$
$$9 + 1 = 10$$

So if we have these numbers to add up –

$$
\begin{array}{c}
9 \\ 3 \\ 1 \\ 4 \\ 6 \\ + \ 7 \\ \hline 30
\end{array}
\qquad \text{we can group them into tens –} \qquad
\begin{array}{c}
8 \\ 4 \\ 2 \\ 6 \\ 5 \\ + \ 5 \\ \hline 30
\end{array}
$$

Now try these
Group these into 10s (using the table if it helps). The answers are on page 92.

5

(a)	(b)	(c)
5	1	7
8	8	5
5	9	6
+ 2	6	3
	2	4
	+ 4	5
		1
		+ 9

LOOKING AT NUMBER

You don't just have to look for pairs of numbers that add up to 10, you could look for bigger groups such as 2, 3 and 5. The more you practise this the easier it becomes to spot numbers that add up to 10.

The same tip can be used with bigger numers, but remember to carry figures over to the next column. The figures carried over are set in smaller type, but you do not have to write them down:

```
    145              1  4  5
    987              9  8  7
    265              2  6  5
+   823        +     8  2  3
                  ₂  ₂  ₂
              ─────────────
                  2  2  2  0
```

Now try these

Remember to add on to the next column any figures you carry over. The answers are on page 92.

	cm.		cm.		miles
6 (a)	**129**	(b)	**326**	(c)	**156**
	378		272		123
	932		637		842
+	781		852		56
			750		42
		+	483	+	161

SIDENUMBER

The metric system has a revolutionary history! After the French Revolution in 1789 it was decided to make a standard system of measurements, based on our counting system of 10. Before then all sorts of measurements were used, which varied from place to place.

This revolutionary attempt at making an easily understood system of measurement is now used in most countries of the world. Britain is steadily and slowly moving towards metrication.

Cash and Carry

Money makes the world go round. We all need to be able to buy and sell and, more importantly, make sure that we are paying (or charging) the right amount. In this section we'll look at some basic ways that can help us.

We often do maths without realising it. For instance, every time we check our change we are doing maths. We probably wouldn't write it down as a 'sum', but most of the maths we do is in our head anyway.

LOOKING AT NUMBER

When we go into a supermarket or a shop and buy some goods we often *guess* roughly how much it will come to. This is called *estimating* and it's a big part of maths.

Try answering this in your head:
If you went into a shop and bought

a newspaper for 30p

a bag of sweets for 48p

and a pen for 14p

would you be able to get them if you had a £5 note? Have a guess.

Would you be able to get them if you only had a £1 coin? Again, have a guess.

When you are doing rough calculations it helps to round the numbers. Rounding means changing a number to one that is about the same, but is easier to work with. We can see how this works with the above prices. Most people find it easier to count in tens and so these prices could be changed to round ten pences. It is usually best to round on the 'safe side'. When you want to be sure that you have enough money to buy what you want you can be safe by rounding the prices up to the next ten pence. With the 48p bag of sweets we would say 48p is more than 40p but less than 50p, so we would call it 50p. The 14p for the pen is between 10p and 20p, so to be safe we would call it 20p. Roughly the newspaper, sweets and pen would be 30p, 50p and 20p which is £1 altogether.

If you want to check the answer exactly, set it out like this and work it out:

$$\begin{array}{r} 30 \\ 48 \\ + \ 14 \\ \scriptstyle 1 \\ \hline 92 \end{array}$$

The answer is 92p, so it is actually less than £1.00.

Sometimes prices are set to try and fool us. Very often we find amounts ending in 99p.

even

When we see signs like these we need to realise that 99p is almost £1. We can make the price simpler by thinking of the 99p as £1 and adding it on. £4.99 would be £5, £4 + £1. This is called rounding up. The shopkeeper would like us to forget the 99p, and think of £4.99 as £4. If we were going to buy 3 items priced at £1.99, £2.99 and £4.99, the shopkeeper would like us to think of the prices as £1, £2 and £4, making £7 altogether. If we round up the prices we can see that they are nearly £2, £3 and £5, or almost £10.

Now try these
Round up these amounts to the nearest pound. The answers are on page 92.

1 (a) **£3.99**

(b) **£49.99**

(c) **£19.99**

(d) **£99.99**

LOOKING AT NUMBER
You are giving someone change. The goods come to £3.37 and you are given a £10 note. What's the easiest and quickest way to do this? It's not to write it down as a sum like this –

$$10.00$$
$$-\ \ \ 3.37$$

– or even to get out a calculator.
The best way is to *count on*, that means count out the change. You've probably often seen this done in shops.

£3.37 and 1p	makes £3.38	①p
£3.38 and 2p	makes £3.40	②p
£3.40 and 10p	makes £3.50	⑩p
£3.50 and 50p	makes £4.00	㊿p
£4.00 and £1	makes £5.00	£1
£5.00 and £5	makes £10.00	£5

You can check your change in the same way – count on the change you're given to the price of the goods. The total *should* come to what you gave.

To take one number from another on paper we write the numbers one under the other in columns, just as when adding numbers. It is important to write them the right way round. We always take the bottom number away from the top one.

$$93$$
$$-\ \ 41$$

The – means take away.

We start with the right-hand column and say: what is 3 take away 1? If you start with 3 and take 1 away you are left with 2. We then move on to the tens column and do 9 take away 4 and get the answer 5.

$$
\begin{array}{r}
93 \\
-\ 41 \\
\hline
52
\end{array}
$$

If we have to take away 0 in one column we are left with the top number in that column:

$$
\begin{array}{r}
37 \\
-\ 20 \\
\hline
17
\end{array}
$$
When we take 0 from 7 we still have 7.

If the bottom number in one column is the same as the top number we will be left with 0.

$$
\begin{array}{r}
65 \\
-\ 25 \\
\hline
40
\end{array}
$$
When we take 5 from 5 we are left with 0.

Now try these
The answers are on page 92.

2 (a) $\begin{array}{r} 35 \\ -\ 14 \\ \hline \end{array}$ (b) $\begin{array}{r} 62 \\ -\ 21 \\ \hline \end{array}$

(c) $\begin{array}{r} 57 \\ -\ 37 \\ \hline \end{array}$ (d) $\begin{array}{r} 48 \\ -\ 24 \\ \hline \end{array}$

There is a problem if we have a column where the number to be taken away (subtracted) is bigger than the number we are taking from. Here are two ways to do this, which both get the same answer. You might have a slightly different way of doing it. It doesn't matter as long as you can get the right answer. If one of these two ways seems familiar then practise with that one and ignore the other. If you don't know which way you find easiest, practise the first way (or a way of your own – there are many other ways).

Why are there different ways? It is just a matter of fashion. At one time everyone was taught the second way at school, and then the fashion changed to the first way.

FIRST WAY

$$
\begin{array}{r}
63 \\
-\ 16 \\
\hline
\end{array}
$$

$$
\begin{array}{r}
{}^{5\,1}63 \\
-\ 16 \\
\hline
\end{array}
$$
We can't take 6 from 3 so take one of the six tens to leave 5 tens and add the one ten to 3 to make 13.

$$
\begin{array}{r}
{}^{5\,1}63 \\
-\ 16 \\
\hline
7
\end{array}
$$
Take the 6 from 13 and it leaves us with 7.

$$
\begin{array}{r}
{}^{5\,1}63 \\
-\ 16 \\
\hline
47
\end{array}
$$
Take the 1 ten from the 5 tens.

SECOND WAY

$$
\begin{array}{r}
63 \\
-\ 16 \\
\hline
\end{array}
$$

$$
\begin{array}{r}
{}^{1}63 \\
-\ 16 \\
\hline
\end{array}
$$
We can't take 6 from 3 so we borrow ten.

$$
\begin{array}{r}
{}^{1}63 \\
-\ 16 \\
\hline
7
\end{array}
$$
That makes 13 and leaves us with 7. But when we borrow ten we have to pay back ten.

$$
\begin{array}{r}
{}^{1}63 \\
-\ 16 \\
{}_{1}\ \\
\hline
47
\end{array}
$$
The 1 ten and the 1 ten we borrowed make 2 tens. Take them from the 6 tens and we are left with 4 tens.

You may find it easier to do the sum with coins. You will need six 10p pieces and thirteen pennies.

Count out 63p using six 10p pieces and three pennies.

To find how much you will have left you need to take away 16p (one 10p piece and six pennies).

FIRST WAY
You need to take away six pennies, but you only have three. To get round this, take one of the 10p pieces and change it into ten pennies. You then have five 10p pieces and thirteen pennies.

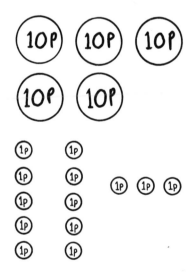

Now take six pennies away to leave seven pennies and five 10p pieces.

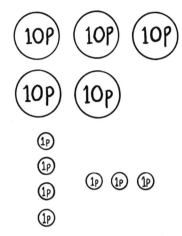

Now you need to take away one 10p piece: this leaves four 10p pieces and seven pennies.

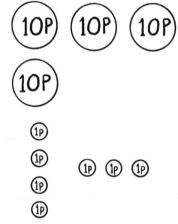

SECOND WAY
The problem is again to take away six pennies when you only have three. This time you borrow ten pennies so that you have six 10p pieces and thirteen pennies.

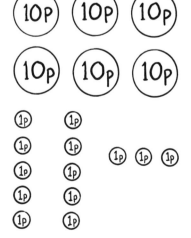

You can now take away the six pennies to leave six 10p pieces and seven pennies.

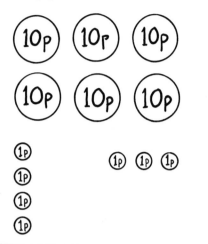

You now need to take away one 10p piece, but you also borrowed 10p, so that is two 10p pieces to take away. That leaves four 10p pieces and seven pennies.

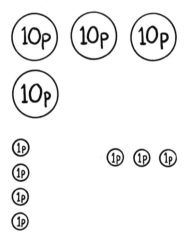

TIP

You can always check this type of calculation (subtraction) – and not just with a calculator or by asking someone!

In the sum with the coins, you started with 63p and split it into two parts, the 16p that you took away and the 47p that was left. If you put these two parts back together again (added them) you get back to the 63p you started with.

63 take away 16 is 47
(often written as

$$\begin{array}{r} 63 \\ -\ 16 \\ \hline 47 \end{array}$$

or **63 – 16 = 47**)

But if we add 47 and 16 we get 63 (often written as

$$\begin{array}{rr} 47 & or\quad 16 \\ +\ 16 & +\ 47 \\ \hline 63 & 63 \end{array}$$

or **47 + 16 = 63**

or **16 + 47 = 63**)

So to check, add what is left to the amount you took away. If you get what you started with the chances are you're right.

Now try these (use coins as well if it helps)
Before you look at the answers on page 92 check your answers by adding.

3   (a) **85**   (b) **47**
     – 39   **– 18**

     (c) **64**   (d) **53**
     – 57   **– 38**

SIDENUMBER

Our way of working with numbers is very much based on the number 10. This is the decimal system; there is more about it in section 5. But this wasn't always so, of course. Decimal money was introduced in the UK in 1971. Before that there had been pounds, shillings and pence.

But the benefits of the decimal system for money had been known for many years before 1971. In Victorian times the British government tried to introduce decimal coins but the idea took a hundred years to catch on. The coin they introduced was called a florin, or 2 shillings – one tenth of a pound. It became our present 10p with the same shape and size of the modern coin. Well, that is until 1992, when the old 10p will be replaced by smaller coins.

It's a Cover Up

We sometimes need to find out how big an area is. If we want to carpet a room, plaster a wall or sow a lawn, we need to work out the area that needs covering. This section looks at ways of working this out.

LOOKING AT NUMBER

Suppose we had a room which we wanted to carpet with carpet tiles. We could guess how many tiles we need, but we would be unlikely to get it right. A better way would be to draw a picture of the floor and draw the carpet tiles on the picture. If the room was 4 metres one way and 3 metres the other, and the tiles were 1 metre by 1 metre, the picture would look like this:

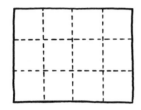

To find the number of carpet tiles we needed we could count them 1, 2, 3, 4, 5, 6, 7, 8, 9, 10, 11, 12. Or we could add them going across – 3 + 3 + 3 + 3 = 12. Or we could add them going from top to bottom

$$\begin{array}{r} 4 \\ 4 \\ + \quad 4 \\ \hline 12 \end{array}$$

We measure area in squares. Each square that is a metre by a metre is called a square metre. Since each carpet tile is 1 square metre we would need 12 square metres of tiles altogether.

The normal way to work out area is to use multiplication. Multiplying means adding on the same number several times.

To show that we want to multiply we use a ×. If we write 4 × 3 that means 4 multiplied by 3. Another way of saying it is 4 times 3. 4 × 3 means 4 added together 3 times, so 4 × 3 is the same as 4 + 4 + 4.

In our room we could multiply the length by the width. In this case it would be

4 metres × 3 metres = 12 square metres

Or we could multiply the width by the length, which will give the same answer:

3 metres × 4 metres = 12 square metres

Remember, multiplying is just a quick way to count or add up equal numbers.

To multiply we need to know how many one number times another is. One way is to know your 'times tables', which are lists of one number times another. Many people will have learnt them off by heart:

One times two is two

Two times two is four

Three times two is six

Four times two is eight

and so on.

If you never learnt times tables, or have difficulty remembering them, use the square on the next page to help you.

This is a multiplication square and is just another way of writing down 'times tables'.

If you want to know what 6 × 7 is, you look *across* the row with 6 at the side until

1	2	3	4	5	6	7	8	9	10
2	4	6	8	10	12	14	16	18	20
3	6	9	12	15	18	21	24	27	30
4	8	12	16	20	24	28	32	36	40
5	10	15	20	25	30	35	40	45	50
6	12	18	24	30	36	42	48	54	60
7	14	21	28	35	42	49	56	63	70
8	16	24	32	40	48	56	64	72	80
9	18	27	36	45	54	63	72	81	90
10	20	30	40	50	60	70	80	90	100

you reach the column with 7 at the top and you find the number 42, which is the answer. To find 4×8, look across the 4 row and down the 8 column and you find the answer 32. You get the same answer if you look across the 8 row and down the 4 column. This is because 4×8 is the same as 8×4 – they are both 32.

Now try these
Do these multiplications, using the table if you need to. The answers are on page 92.

1

    (a) 5×3

    (b) 6×8

    (c) 7×4

    (d) 4×9

TIP
To multiply a number by 10 simply add 0 to the end. For example, you can see from the number square

$$4 \times 10 = 40$$

$$7 \times 10 = 70$$

This is true for all numbers no matter how big. For example

$$50 \times 10 = 500$$

$$71\,956 \times 10 = 719\,560$$

LOOKING AT NUMBER

Often we need to know the area of a shape that we can't fill exactly with whole squares.

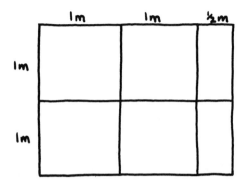

Here we have four whole squares but we've also got two half-squares left over. These two half-squares make one full square, giving us five squares altogether. If each square was 1 metre by 1 metre they would be 1 square metre each, so the total area would be 5 square metres (sometimes written as 5 sq.m. or 5m.^2).

Different shapes can have the same area. If you take a piece of paper

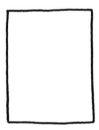

cut it in half across the middle and put the pieces side by side

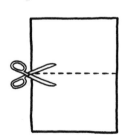

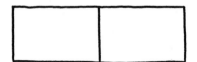

you get a different shape, but you still have the same area because you have the same amount of paper.

Now try these
The answers are on page 92.
This bathroom wall needs retiling.

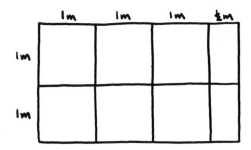

2 (a) How many square metres of tiles would it need?
  (b) The tiles cost £10 for a square metre, what would the total cost be?

This lawn needs re-seeding

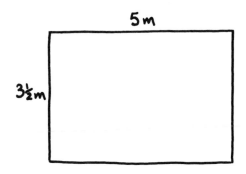

(c) How many square metres is it?

LOOKING AT NUMBER
Sometimes we need to be able to find out the area of shapes which aren't so simple. Sometimes you can split the shape up into simpler parts. With this L-shaped room you could split it up like this:

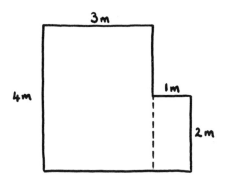

and work out the areas of the parts.

$$4 \times 3 = 12 \text{ sq. m.}$$

$$2 \times 1 = 2 \text{ sq. m.}$$

Then add them together

$$12 + 2 = 14 \text{ sq. m.}$$

If the shape is very irregular try putting a grid over it and then count (or multiply if that's easier) the full squares and estimate 'the bits' left over.

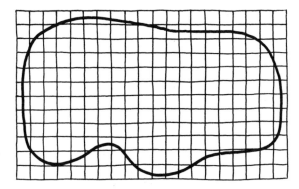

LOOKING AT NUMBER
The multiplication square doesn't tell us the answer when we want to multiply a number bigger than 10. To do this we have to split the number into parts. To multiply 21 by 4 we can say that 21 is actually 2 tens and 1 extra. Start with the 1 and multiply that by 4, which gives 4. Go on to the 2 tens: 2 multiplied by 4 is 8, which is 8 tens. 8 tens and 4 extra is 84 altogether. We write this sum like this:

$$\begin{array}{r} 21 \\ \times\ 4 \\ \hline 84 \end{array}$$

We would work out 25×3 like this

$$\begin{array}{r} 25 \\ \times\ 3 \\ \hline 75 \end{array}$$

When we multiply the 5 and the 3 we get 15. The 15 is 1 ten and 5 extra, so we write down the 5 and add the 1 ten on to the 6 tens we get when we multiply the 2 and the 3, giving 7 tens altogether.

SIDENUMBER

Areas of land are often measured in acres. An acre was originally the amount of land a team of oxen could plough in a day, but nowadays not many people in this country use oxen! An acre is now officially set at 4840 square yards. This may be difficult to imagine but roughly – it's about 50 paces one way and 100 the other, if the shape is like this

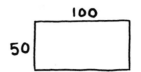

or 25 by 200, if it's like this.

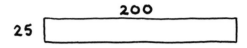

The football pitch at Wembley is about 2 acres.

I'll Cut and You Choose

Counting small numbers is no problem. But if we need to count larger numbers it is often better to group the items together.

Goods are often grouped together in batches. It is far easier to count the number of batches than to count each item one by one. Different goods can be packed in different-sized groups.

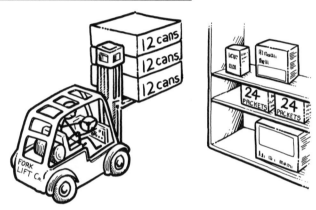

LOOKING AT NUMBER

When we multiply we are just adding or counting very quickly. Suppose we had 8 crates of lemonade to deliver and each crate held 6 bottles, we can find how many bottles that would be by multiplying. Check the multiplication square in section 3. If we look across the 8 row and down the 6 column they meet at 48. We would have 48 bottles.

What if we had to do the opposite, share out or divide the 48 bottles into crates each holding 6 bottles. How many crates would that be?

We can use the same table, but this time we look down the 6 column until we come to 48 and then look back along the row to see which one we are on, and see the answer 8.

Division (the sign for it is ÷) is simply sharing out.

Imagine we had 12 cabbages and we wanted to plant them in rows, we could plant them

like this

or like this

Depending on how we look at it, we have divided them into 3 rows each of 4 cabbages, or 4 rows each with 3 cabbages. Either way we still have 12 cabbages because 3 × 4 and 4 × 3 both give the answer 12.

Division is the opposite of multiplication. When we multiply we take several equal-sized groups and put them together as one large group. When we divide we take one large group and split it into several equal-sized groups.

You can see how division works by sharing out some real things, such as matches or coins.

If you want to divide 10 by 5 it simply means starting with 10 items and counting them out into 5 equal piles.

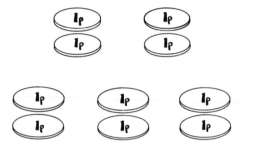

Now try these
Use the multiplication table in section 3, or some matches or coins, if it helps you. The answers are on page 92.

1 (a) 24 divided by 6
   (b) 42 divided by 7
   (c) 72 divided by 9

LOOKING AT NUMBER
There are different ways of setting out division sums. It doesn't really matter which way you write it down, although for more complicated division sums the first way is more convenient.

One way is

$$6 \overline{)24}^{\,4}$$

This means: how many 6s are there in 24? There are 4.

Another way is

$$6 \underline{)24}$$
$$4$$

This still means: how many 6s are there in 24? The answer is still 4.

Another way is

$$24 \div 6 = 4$$

And this still means: how many 6s are there in 24? The answer is still 4.

Yet another way is

$$\frac{24}{6} = 4$$

Yet again the answer is 4.

Now try these
The answers are on page 92.

2   (a) **18 ÷ 9**
            (b) **30 ÷ 5**
            (c) **56 ÷ 8**

LOOKING AT NUMBER
Division can often help us to spot the best buy.

The Tesburys Fish Fingas are the cheapest but are they the best bargain? Division can help.

The Tesburys Fish Fingas cost 40p and there are 5 fish fingers. So each fish finger costs 40 divided by 5 or (same thing) 40 ÷ 5 = 8. Each fish finger costs 8p.

The Sainsco Fishy Fingers cost 70p and there are 10 fish fingers. So each fish finger costs 70 divided by 10 or 70 ÷ 10 = 7. Each fish finger in this packet costs only 7p.

Remember, it isn't a bargain to buy in bulk if you can't afford it, or if you don't need the amount you've bought.

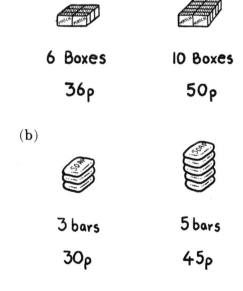

TIP
To divide any whole number ending in 0 by ten cross off the last 0.

$$40 \div 10 = 4$$

$$\pounds350 \div 10 = \pounds35$$

To divide money by 10 move the dot separating the pounds and the pennies one place to the left, and drop off the last number.

$$\pounds25.50 \div 10 = \pounds2.55$$

$$\pounds141.30 \div 10 = \pounds14.13$$

Now try these
Use division to find which is the best buy. The answers are on page 92.

3 (a)

6 Boxes   10 Boxes

36p   50p

(b)

3 bars   5 bars

30p   45p

LOOKING AT NUMBER
Sometimes, if you don't need an exact answer, it is easier to round numbers before you divide, as for rounding prices up for estimating in section 2.

Here we aren't working these out exactly, but estimating which is the best buy.

Only £2.99

Only £11.99

With the tapes at 3 for £2.99 it is easier to call the price £3. We can then see that they are about £1 each, since £3 ÷ 3 = £1.

With the 10 for £11.99 pack we can call the price £12.00. We then divide by 10 by moving the dot, and find that these tapes cost about £1.20 each.

Don't be fooled: sometimes 'best buys' aren't really best buys.

CHEAPO RECORDS
As Sung by Elvis.
only £1.90

DEARA ALBUMS
Sung By Elvis...

SIDENUMBER
We often talk about odd and even numbers. How do we know which are odd and which are even? One way is to write the numbers down in order starting from 1. Every other number is odd and all the rest are even.

Odd	Even
1	
	2
3	
	4
5	
	6
7	
	8
9	
	10
11	
	12

An easier way is to see if the number can be divided exactly by 2. If it can then the number is even, if not it is very odd!

Your Number's Up

So far we have looked at numbers as ways of helping us measure. We saw how to measure length in section 1, money in section 2, area in section 3, and quantities in section 4.

The beauty of numbers is that they can be used for anything and everything. You can use them for money, for cricket scores, or even to describe the number of hairs in a toupee!

I'd Like one with 36 thousand hairs please.

In this section we look at the basic ways in which numbers can be seen and handled.

LOOKING AT NUMBER
Number work is simply counting, but to do everything just by counting would be very slow and we would be likely to make mistakes. Adding is just a quick way of counting up. We have a variety of ways of indicating addition. We could say:

5 add 4

or **5 plus 4**

or **5 and 4**

or **5 + 4**

They all mean the same thing.

So far we have only looked at adding with fairly small numbers, but the same methods work just as well for large numbers. It just takes longer. It is merely a matter of writing the numbers down in columns and then adding up column by column starting at the right-hand side. We don't often have to deal with large numbers in our personal life, but in many jobs it is sometimes necessary.

If the daily takings in a shop were as follows:

Monday	**£387**
Tuesday	**£326**
Wednesday	**£491**
Thursday	**£410**
Friday	**£623**
Saturday	**£1052**

we could work out the weekly takings by adding them all together:

```
           £
         387
         326
         491
         410
         623
    +   1052
         2 2 1
       _____
        3289
```

Start at the right-hand side, working column by column and carrying over into the next column where necessary. The numbers to be carried have been added in smaller type at the bottom of each column. Where and how you write them is up to you, as long as it is clear what they are.

Now try these
Add up these numbers. The answers are on page 92.

1 (a)
```
      386
      727
  +   193
    _____
```
(b)
```
      275
      285
      614
  +   326
    _____
```

(c)
```
     1243
     2793
  +  4288
    _____
```
(d)
```
     6438
     7231
     4625
     5622
  +  2383
    _____
```

LOOKING AT NUMBER
Subtraction is the quick way to count down, and again there are several ways of showing this. We could say:

8 subtract 4

or **8 take away 4**

or **8 minus 4**

or **8 – 4**

They all mean the same thing.
If you want to do subtraction with large numbers, work column by column, as with small numbers. Start at the right-hand side:

```
     2586
  -  1432
    _____
     1154
```

When it is necessary to 'borrow' you can use the same method that you used for small numbers. If you use the first method of subtraction shown in section 2 you sometimes have to 'borrow' across several columns.

```
      200
  -   136
    _____
```

Starting from the right-hand column we want to take 6 away from 0, which we can't do, so we would normally 'borrow' from the next column to the left. In this sum, however, there is nothing to 'borrow' in that column. To get something in this column we need to 'borrow' from the next one along.

```
   200   becomes    ¹100   (one hundred
-  136            -  136    and ten tens)
  ____             _____
```

We now have something we can 'borrow':

```
  ¹100   becomes    ¹190   (one hundred,
-  136            -  136    nine tens and
  ____             _____    ten)
```

Now we can do the sum:

$$\begin{array}{r} \overset{1}{1}90 \\ - 136 \\ \hline 64 \end{array}$$

Now try these
Take away the bottom number from the top one. The answers are on page 92.

2 (a) $\begin{array}{r} 357 \\ -143 \\ \hline \end{array}$ (b) $\begin{array}{r} 642 \\ -264 \\ \hline \end{array}$ (c) $\begin{array}{r} 1585 \\ -1342 \\ \hline \end{array}$

(d) $\begin{array}{r} 3542 \\ -1863 \\ \hline \end{array}$ (e) $\begin{array}{r} 5000 \\ -2654 \\ \hline \end{array}$

LOOKING AT NUMBER
When we want to add the same number several times over we can use multiplication as a shortcut.

We usually say **5 times 6**

or write **5 × 6**

sometimes we say **5 multiplied by 6**

or **5 lots of 6**

or even **the product of 5 and 6**
(no one really says this, but it is sometimes written in textbooks)

Several things about multiplication are worth knowing. The first, about multiplying by 10, was mentioned in section 3. The same tip applies to multiplying by 100 or 1000 or any number which is 1 followed by 0s. When multiplying by a number of this sort, simply add the same number of 0s to the original number. So to multiply by 100, add 2 0s; to multiply by 1000, add 3 0s, and so on.

$12 \times 100 = 1200$

$4356 \times 100 = 435\,600$

$12 \times 1000 = 12\,000$

$4356 \times 1000 = 4\,356\,000$

Multiplying can be done in several steps. Multiplying by 20 is the same as multiplying by 2 and then by 10, since $2 \times 10 = 20$. To work out 35×20 we could work out 35×2 and then multiply the answer by 10.

$35 \times 2 = 70$

$70 \times 10 = 700$ (just add a 0)

so $35 \times 20 = 700$

Since multiplying is the same as repeatedly adding on the same number, you can split the multiplication up into simpler parts. To multiply a number by 12 you can first multiply it by 2 and then by 10. Then if you add the answers together you would get the answer for multiplying by 12, since $2 + 10 = 12$.

$24 \times 2 = 48$ and $24 \times 10 = 240$

so $24 \times 12 = 48 + 240 = 288$

You can use all of these points together to multiply any numbers, however large. If, for example, you wanted to multiply 45×36 you could do it as follows:

45×36 is the same as 45×6 added to

45×30

You can work out 45×6 like this:

$$\begin{array}{r} 45 \\ \times 6 \\ {\scriptstyle 3} \\ \hline 270 \end{array}$$

You can work out 45×30 by saying it is 45×3 and then $\times 10$:

$$
\begin{array}{r}
45 \\
\times \quad {}_1 3 \\
\hline
135
\end{array}
$$

and $\quad$ **$135 \times 10 = 1350$**

so $\quad$ **$45 \times 36 = 270 + 1350 = 1620$**

This method of multiplying large numbers is called long multiplication. The normal way of writing a long multiplication sum is slightly different to the way it has just been done. This is the same sum written out differently:

$$
\begin{array}{r}
45 \\
\times \quad 36 \\
\hline
\end{array}
$$

The first step is to work out 45×6 and write the answer below the line:

$$
\begin{array}{r}
45 \\
\times \quad 36 \\
\hline
270
\end{array}
$$

Next work out 45×30. You do this by first writing the 0 to multiply by 10:

$$
\begin{array}{r}
45 \\
\times \quad 36 \\
\hline
270 \\
0
\end{array}
$$

Then work out 45×3 and write the answer alongside the 0:

$$
\begin{array}{r}
45 \\
\times \quad 36 \\
\hline
270 \\
1350
\end{array}
$$

Finally, add the answers together:

$$
\begin{array}{r}
45 \\
\times \quad 36 \\
\hline
270 \\
+ \quad 1350 \\
\hline
1620
\end{array}
$$

Some people would do this sum by working out 45×30 first and then 45×6. It doesn't make any difference which way round you do it. You should still get the same answer.

With even larger numbers you use the same method of splitting the sum up into simpler parts; it is just more long-winded.

To work out 136×124, split it into 136×4, 136×20 and 136×100.

Now try these
Work out these long multiplications. The answers are on page 92.

3 (a) $\begin{array}{r} 36 \\ \times \quad 15 \\ \hline \end{array}$ $\qquad$ (b) $\begin{array}{r} 48 \\ \times \quad 26 \\ \hline \end{array}$

(c) $\begin{array}{r} 51 \\ \times \quad 32 \\ \hline \end{array}$ $\qquad$ (d) $\begin{array}{r} 385 \\ \times \quad 176 \\ \hline \end{array}$

LOOKING AT NUMBER
Division is a way of subtracting the same amount several times over, but it is easiest to think of it as sharing into equal amounts. We could say:

15 divided by 3

or $\quad$ **15 shared by 3**

or $\quad$ **$15 \div 3$**

In section 4 we looked at how to divide by using the multiplication square. The problem with this method is that the largest number in the square is 100. We need a way of dealing with dividing or sharing larger numbers. We can do this by splitting the sum into parts.

To share 639 into 3 parts we start with the hundreds.

There are 6 hundreds. If we share 6 into 3 parts we get 2 in each part (6 ÷ 3 = 2), so each part will have 2 hundreds.

Now move on to the tens. There are 3 tens, so if we share them into 3 parts we get 1 ten in each (3 ÷ 3 = 1).

Finally, we move on to the ones. There are 9 ones. When we split them into 3 we have 3 in each part (9 ÷ 3 = 3).

So if we share 639 into 3 parts each part will have 2 hundreds, 1 ten and 3 ones, or 213. We write this sum as follows:

3 $\overline{639}$ This means 639 divided by 3.

We start with the hundreds

$$\begin{array}{r} 2 \\ 3\overline{)639} \end{array}$$

then do the tens

$$\begin{array}{r} 21 \\ 3\overline{)639} \end{array}$$

then the ones

$$\begin{array}{r} 213 \\ 3\overline{)639} \end{array}$$

The number of hundreds, tens and ones in the answer are written above the hundreds, tens and ones in the number being divided.

Now try these
Write these out in the same way as the sum above. You may need to use the multiplication table in section 3. The answers are on page 92.

4 (a) **248 ÷ 2**   (b) **369 ÷ 3**
    (c) **484 ÷ 4**   (d) **864 ÷ 2**

LOOKING AT NUMBER
Sometimes you can't share equally. If we want to divide 736 by 2 we can start with

the 7 hundreds, but when we look in the 2 column of the multiplication table, 7 is not there. The way we deal with this is to find the highest number that does appear that is less than 7. This is 6, and 6 ÷ 2 is 3. So if we share 7 hundreds into 2 parts we get 3 hundreds in each, but will have 1 hundred left over. To share out this extra hundred we split it into 10 tens and add it to the 3 tens we already have, giving 13 tens altogether. We then share the 13 tens. The 13 tens cannot be shared equally, but we can share 12 tens into 6 lots of 2 and have 1 ten left over. Change the extra ten into 10 ones to go with the 6 ones we already have to give 16 altogether. This can be shared into 8 lots of 2. So when we divide 736 into 2 parts we end up with 3 hundreds, 6 tens and 8 ones in each part, or 368.

We write this as $\begin{array}{r} 368 \\ 2\overline{)7^13^16} \end{array}$

The small numbers show the extra ones from one column that are carried over to the next column.

Sometimes there is not enough in a column to share out. In this case, write a 0 to show there is nothing in each part:

$$\begin{array}{r} 102 \\ 6\overline{)612} \end{array}$$

When we come to the tens column there is only 1 ten. When we try and divide 1 ten into 6 parts we get 0 in each part with 1 ten left over which we carry over to the 2 ones to give 12 ones.

Now try these
The answers are on page 93.

5 (a) **725 ÷ 5**   (b) **714 ÷ 6**
    (c) **840 ÷ 7**   (d) **256 ÷ 8**

You may have found long multiplication complicated. Long division, which is dividing large numbers by large numbers, is even more difficult.

TIP
If you want to do a division sum which involves large numbers, use a calculator! Calculators enable us to do long and complicated sums without effort. Of course it is important that you press the right keys. This may sound obvious, but it is very easy to press a button by mistake, which will give the wrong answer.

Now try these
Only do these if you have a calculator. The answers are on page 93.

6 (a) **925 ÷ 37** (b) **22 843 ÷ 53**
(c) **83 354 ÷ 142** (d) **538 704 ÷ 432**

SIDENUMBER
We write our numbers using place value because this makes it much easier to do sums. The Romans had a different system of numbers, which did not use place value. You sometimes see Roman numbers on clocks or at the end of a television programme, when the date is shown.

The Roman system uses letters:

I   means 1
V   means 5
X   means 10
L   means 50
C   means 100
D   means 500
M   means 1000

Numbers are made up by combining these letters. When a smaller number is written after a larger one it is added to the larger one. So **VI** means 5 + 1, or 6. If the smaller number is written first it is taken away from the larger one, therefore **IV** means 5 – 1, or 4.

The Time of Day

At one time people got up in the morning when the sun rose and went to bed when it got dark. They didn't need to be able to tell the exact time, and could always count the chimes from the church clock. These days we increasingly need to be able to tell the time and measure time.

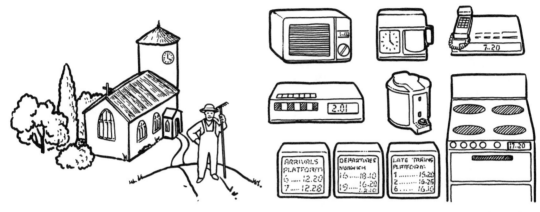

LOOKING AT NUMBER

To measure time we split the day up into two lots of 12 hours (2 × 12 = 24 hours altogether). The first group of 12 hours starts at midnight and the second 12 hours start at midday. Each hour is split into 60 minutes. When we talk about times in the morning we say how many hours and minutes it is since midnight. In the morning eight-thirty (written in numbers as 8:30) means eight hours and thirty minutes since midnight. In the afternoon we say how many hours and minutes it is since midday: five or five o'clock means five hours and no minutes since midday. We often say o'clock when it is a whole number of hours and no minutes. When we talk about a time it is usually fairly obvious whether we mean morning or afternoon, but sometimes there could be confusion. To make it clear which we mean we call times in the morning a.m., and times in the afternoon p.m.

If someone says they finish work at five-thirty we normally assume that they mean in the afternoon. If, however, we want someone to meet us at the station at 9:30 it would be best if we say whether we

mean 9:30 a.m. or 9:30 p.m., otherwise we might have a long wait!

There are two different sorts of clocks and watches nowadays, digital and analogue (pronounced anna-log). Most people find digital watches easier to use since they show the time as a number.

This is six-twenty, which means six hours and twenty minutes since midnight or midday. The people who make digital watches normally assume that we know whether it is morning or afternoon.

Analogue clocks and watches usually have a round or square 'face', sometimes with numbers on it and two 'hands'. The hands point to the numbers on the face to show the time; a small hand points to the hours, and a big one points to the minutes. A clock face is really two faces placed over one another. One for the small hand shows the hours, and the other for the big hand shows the minutes.

2 o'clock

10 minutes past the hour

When a face doesn't have any numbers on it, we have to imagine that the numbers are where they are in the picture.

The hands on an analogue clock turn round slowly, so they do not always point directly at a number. To read the hours on an analogue clock, work out which number the small hand was last pointing at. Then read the minutes by seeing where the big hand is pointing.

The time here is one-twenty. The small hand is between the 1 and the 2, so the last number it pointed at was 1.

The time here is five-fifty. Although the small hand is almost pointing to 6, it is not quite there yet.

Now try these
Write down the times on these clocks in words. The answers are on page 93.

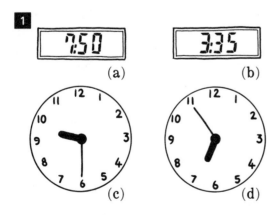

1

(a)

(b)

(c)

(d)

LOOKING AT NUMBER

So far we have looked at telling the time by saying the number of hours and then the number of minutes, so six-ten means six hours and ten minutes from either midnight or midday.

Another way to talk about time is to say how many minutes it is since the last hour or how many minutes until the next hour: we use 'past' and 'to' for this.

Twenty-five past four means it is twenty-five minutes since four o'clock. This is the same as four twenty-five.

Ten to seven means in another ten minutes it will be seven o'clock. There are sixty minutes in an hour, so when it is ten minutes to the next hour it is fifty minutes since the last hour. So ten to seven is the same as six-fifty.

With this way of talking about time we only say how many minutes 'to' the next hour if it is less than thirty. We say twenty-five minutes past, rather than thirty-five minutes to.

For fifteen, thirty and forty-five minutes we sometimes use the following words:

**15 minutes (or mins) is
a quarter past the hour**

**30 minutes is
half past the hour**

**45 minutes is
a quarter to the next hour**

Here are some times shown on digital and analogue clocks, with the different ways of saying them.

is the same as

we could say seven
or seven o'clock

is the same as

we could say two-thirty
or half-past two

is the same as we could say nine-fifteen or a quarter-past nine

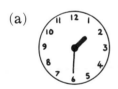 is the same as we could say one forty-five or a quarter to two

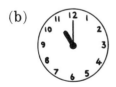 is the same as we could say ten-twenty or twenty-past ten

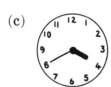 is the same as we could say eight-fifty or ten to nine

Now try these
Try reading these times. The answers are on page 93, in both words and figures.

2 (a)

(b)

(c)

TIP
The best way to tell the time is to use a real clock or watch. Keep moving the hands and read off the hours first – 1 to 12; then move the minute hand 5 minutes at a time. You could check what you are doing with someone else.

LOOKING AT NUMBER
We often need to do sums with time. One example is when we are cooking:

This is really an addition problem.
Check your watch or clock and add on 20 minutes.

$$\begin{array}{r} 8{:}15 \\ +\quad 20 \\ \hline 8{:}35 \end{array}$$

At other times you need to take away one time from another. If the bus leaves at 7:30 and it takes 10 minutes to get to the bus stop what time do you need to leave home?

$$\begin{array}{r} 7{:}30 \\ -\quad 10 \\ \hline 7{:}20 \end{array}$$

Now try these
You might want to use a clock or watch to help you to work these out. The answers are on page 93.

3 (a) Add on 15 minutes to ten-past ten.
(b) Add on half an hour to five-past six.
(c) Take away 5 minutes from 11:05.
(d) Take away a quarter of an hour from 8:30.

LOOKING AT NUMBER
Sometimes the sums are more difficult.
If we leave home at 7:45 and our journey takes 1 hour and 30 minutes, when will we arrive?

$$\begin{array}{r} 7{:}45 \\ +\quad 1{:}30 \\ \hline 9{:}15 \end{array}$$

Here we carry an hour when we get 6 lots of ten minutes, because there are 60 and not 100 minutes in an hour.

SIDENUMBER
We are used to measuring time with clocks and watches, but in the past all sorts of instruments were used: marked candles that burned down slowly, sundials that cast shadows, large egg-timers, even clocks worked by water.

Weight for Me

Everything has weight, even a speck of dust. Often we can get by with just knowing the rough or approximate weight – anyone working on a fruit and veg stall soon gets to know what a pound of apples feels like. But sometimes we like to know exactly. The stallholder would soon go broke, or make a fortune, if he or she just estimated all the time! That's why scales were invented.

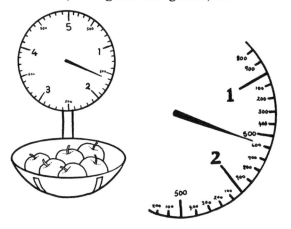

LOOKING AT NUMBER

As we saw in section 1, lengths are measured in both metric and imperial units. Sometimes we might use one system and sometimes the other. The same is true of weight.

Some foods in a supermarket, like tea, coffee, sugar and flour, are weighed in metric (kilograms and grams), and some, like potatoes and fruit, are weighed in imperial (pounds and ounces). It may seem confusing, but usually items that are sold loose, such as fruit, cheese and meat, are weighed in imperial units, and those that are ready-packed are weighed in metric.

There are two common types of displays used on scales. They are similar to those used for measuring time that we looked at in the last section. These are digital scales and analogue scales.

Digital scales normally measure weights in kilograms and grams (there are 1000 grams in a kilogram).

Digital scales which showed

would mean 1 kilogram and 600 grams.

Analogue scales would show the same amount (1 kilogram 600 grams) as

Digital scales, like digital watches, are normally easier to read since they show the weight as a number.

With analogue scales we need to decide which line is nearest to where the hand, or needle, is pointing. If the line has a number against it then that tells the weight, but only some of the lines are numbered. If the needle is pointing to a line without a number, then count round from a numbered line to work out what the weight is.

Now try these
Read these weights in kilograms and grams. The answers are on page 93.

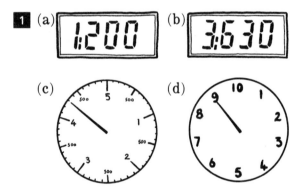

LOOKING AT NUMBER
Small scales that use imperial weights, such as kitchen scales, usually weigh in pounds and ounces. Scales for larger weights, such as bathroom scales, weigh in stones and pounds.

There are 16 ounces (oz.) in 1 pound (lb.)

There are 14 pounds (lb.) in 1 stone (st.)

8 ounces is the same as half a pound

4 ounces is the same as a quarter of a pound

TIP
We don't usually need to know the exact weight of the food we buy. We go by the approximate amount we think we need. The exact weight is only important when we come to pay.

In the kitchen or a supermarket you can get the 'feel' of different weights. Packets of food have the weight printed on them. To see what a kilogram feels like pick up a standard-sized bag of sugar. If you want to see what a pound feels like pick up three large eating apples: they will weigh about a pound. Try to avoid items that have heavy packaging – a 200-gram jar of coffee is a lot heavier than 200 grams!

LOOKING AT NUMBER
It is sometimes useful to be able to change weights from one system to the other. Here is a rough guide to help you:

1 kilogram (or kg.) is just over 2 pounds (or lb.)

30 grams (or g.) are about the same as 1 ounce (or oz.)

1 pound (lb.) is about the same as half a kilogram (kg.)

1 ounce (oz.) is about the same as 30 grams (g.)

So very approximately

1 lb. of cheese is about 500 grams

Half a pound of bacon is about 250 grams

A quarter of a pound of tea is about 125 grams

Now try these
Try to convert these weights from one system to the other using the rough guide given here. The answers are on page 93.

2 (a) **5 kilograms of cement**

    (b) **500 grams of grapes**

    (c) **1 pound of apples**

    (d) **4 ounces of butter**

LOOKING AT NUMBER
Working out sums involving weights is quite straightforward with the metric system. Because this system is based on multiples of ten it fits in with our normal way of working with numbers.

If we want to add 350 grams and 125 grams we can just write it as a normal addition and add up the columns.

$$
\begin{array}{r}
350\text{ g.} \\
+\quad 125\text{ g.} \\
\hline
475\text{ g.}
\end{array}
$$

Subtraction is also the same as it is with normal numbers.

We could use subtraction to find the weight of a dog! The way to do this is to weigh yourself, and then weigh yourself holding the dog. Take one weight away from the other to find the weight of the dog.

Man   **Man and dog**
70 kg.   **80 kg.**

$$
\begin{array}{r}
80\text{ kg.} \\
-\quad 70\text{ kg.} \\
\hline
\text{So the dog is}\quad 10\text{ kg.}
\end{array}
$$

In cooking you may need to multiply or divide weights, as the amounts given in recipes aren't always for the number of people you want to feed.

Here is a list of the ingredients to make enough pastry for 20 mince pies.

40 g. of lard

40 g. of margarine

180 g. of plain flour

A pinch of salt

Cold water

To make 60 mince pies we could either make up three lots of pastry using these amounts, or we could make one lot with three times the ingredients. We can work out what three lots of 40 g. is by multiplying.

$$
\begin{array}{r}
40\text{ g.} \\
\times\quad 3 \\
\hline
120\text{ g.}
\end{array}
$$

To make 5 mince pies we could work out how much we need by dividing by 4

$$
\begin{array}{r}
10\text{ g.} \\
4\,\overline{\big)\,40\text{ g.}}
\end{array}
$$

Now try these
Work out these sums involving weights. The answers are on page 93.

3 (a) **140 g. + 375 g.**

    (b) **970 g. – 665 g.**

    (c) **125 g. × 4**

    (d) **75 g. ÷ 3**

LOOKING AT NUMBER

Doing sums using imperial weights is a lot harder, because they are not based on tens. When you get 16 ounces it is the same as 1 pound, and 14 pounds is the same as 1 stone.

SIDENUMBER

As already pointed out, the metric system is easier to work with because it is based on ten and multiples of ten. A kilogram is 1000 grams (kilo comes from the Greek word for 1000. We also have kilometre = 1000 metres).

Some of the older measures were quite strange and not at all easy to deal with:

16 drams = 1 ounce

2 stones = 1 quarter

4 quarters = 1 hundredweight

And a ton, how heavy is that ? It depends if you mean a long ton or a short ton:

1 long ton = 20 hundredweight

1 short ton = 2000 pounds

Finally, in the metric system we have a tonne which is 1000 kilograms and is very similar in weight to a long ton.

Mine's a Pint

The last section looked at measuring weight, but with some things, particularly liquids and gases, it is usually more convenient to measure how much space they take up. This is called volume. Things that weigh very little can have a large volume. The gas in a gasometer may not weigh much, but it certainly takes up a lot of space.

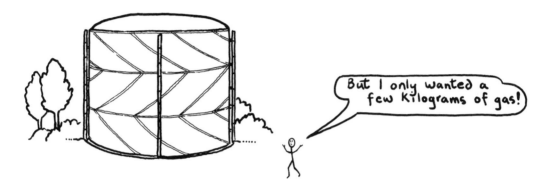

But I only wanted a few kilograms of gas!

LOOKING AT NUMBER

In this country volume is measured, like the other measurements we have looked at, by using both the metric and the imperial systems.

The metric system is, of course, based on ten and multiples of ten. The basic measure is the litre (or l.). This is split into 100 parts called centilitres (cl.) and each centilitre can be split into 10 millilitres (ml.). So there are 1000 millilitres in a litre.

In the imperial system the basic unit is the pint (pt.). There are two other units that are commonly used: the fluid once (fl.oz.), and the gallon (gal.).

20 fluid ounces = 1 pint

8 pints = 1 gallon

TIP

With volume it is very easy to be fooled by the shape of the container. A tall thin bottle often doesn't hold any more than a shorter, slightly fatter bottle. The old-style tall milk bottle looked as if it held more than the new shorter bottles. The only way to be sure is to read the label.

To prove that the shape of the container doesn't alter the volume, find some empty containers, such as drinks cartons and bottles, which are marked with the same volume. Fill one of them up with water and then pour the water from one container to another. As long as you don't spill any, you should see that, although one container may look a lot bigger than the other, both hold the same amount.

LOOKING AT NUMBER

Most things that we buy by volume, such as drinks and petrol, are either prepackaged or delivered by metered pumps. The only time most people need to measure volume for themselves is in the kitchen. Here we generally use a measuring jug, which is a jug with lines marked up the side to show the volume. Simply pour the liquid into the jug until it reaches the required mark.

TIP

As with measures of length and weight it is useful to have a 'feel' for the different units. Most people know what a pint of beer or a pint of milk looks like. To see what a litre looks like go into a supermarket and look at the soft drinks. These are normally sold in one- or two-litre bottles and cartons. A large can of motor oil used to hold a gallon, but nowadays the can is slightly larger, and holds 5 litres. Most wine bottles hold 75 centilitres, and a medicine spoon 5 millilitres.

LOOKING AT NUMBER

Here are some rough guidelines for changing from one system to the other:

1 litre is about one and three-quarter pints

1 centilitre is about a third of a fluid ounce

or

1 fluid ounce is about 3 centilitres

1 pint is about 60 centilitres

1 gallon is about four and a half litres

Now try these
Convert these to imperial. The answers are on page 93.

1 (a) **120 cl.**　(b) **15 cl.**　(c) **9 l.**

Convert these to metric. The answers are on page 93.

2 (a) **10 fl. oz.**　(b) **5 pt.**　(c) **2 gal.**

TIP

Recipes may give measurements in American units, which are different to both imperial and metric. Make sure that you know which units are being used.

LOOKING AT NUMBER

There are other units for measuring volume in both the metric and imperial systems. These are mostly used for measuring gases, empty spaces and solids.

The cubic centimetre (c.c.) is often used for measuring the capacity of car engines. If you imagine a tiny box about the size of a sugar cube, which is 1 cm. along each side, that would have a volume of 1c.c.

A gas meter measures gas in cubic feet. A cubic foot is the volume of a box which is 1 foot along each side.

Ready-mixed concrete is sold in cubic metres. That would be a box 1m. along each side.

Working out a volume in a cubic measure is similar to working out area, but instead of just multiplying length by width, as we would for area, we also have to multiply by height or depth.

To work out the volume of this hole, measure its length, width and depth and then multiply them all together.

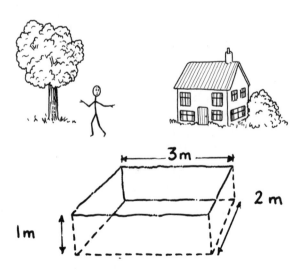

2 metres × 3 metres × 1 metre = 6 cubic metres (usually written as 6 m.³).

It is important when working out a cubic measure to make sure that you measure each length in the same units. If you measure one in centimetres then you must measure the others in centimetres as well.

One thing to note is that although there are only 100 centimetres in a metre there are 1 000 000 (1 million) cubic centimetres in a cubic metre.

Now try these
The answers are on page 93.
3 (a) How many cubic metres of concrete would you need to fill this hole?

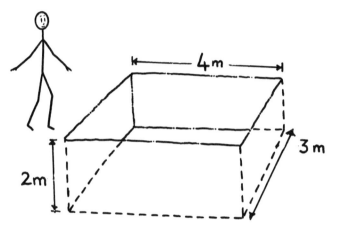

(b) How many cubic metres would you need if it were twice as deep?

LOOKING AT NUMBER
Petrol is measured by volume. Although it is sold in litres nowadays many people still think of petrol in terms of gallons, and many garages still quote a price per gallon.

Cars are often compared to each other in terms of miles per gallon, which is the number of miles they can go on one gallon of petrol.

To work out the fuel consumption of a car, note how much petrol you put in the car, and how many miles you travel using that petrol. Then divide the number of miles travelled by the number of gallons used to give the number of miles per gallon (m.p.g.).

For example, if you travelled 175 miles using 5 gallons of petrol, the m.p.g. is worked out as follows:

$$5 \overline{\smash{\big)}\, 175} = 35$$

So the fuel consumption would be 35 m.p.g.

Now try these
The answers are on page 93.

4 (a) If a car travels 240 miles using 6 gallons of petrol what is the m.p.g.?
   (b) Another car travels 270 miles using 9 gallons. What is its m.p.g.?
   (c) How many gallons would a car that does 32 miles per gallon use to travel 256 miles? (You may need to use a calculator for this.)
   (d) If petrol costs £1.80 a gallon, how much would the journey in question (c) cost?

SIDENUMBER

As we saw earlier, the metric system was introduced after the French Revolution as a way of standardising measurements. The units for length, weight and volume were all related. The metre was originally worked out by measuring the distance from the equator to the North Pole and dividing by 10 million! The litre was set as a volume of a cube with each side being 10 cm. The kilogram was originally the weight of a litre of pure water.

Since then it has been found that the measurement from the equator to the North Pole was actually 2100 metres too short, and so the original basis for the whole system was slightly wrong. Measurements are now set using different methods.

In the imperial system, some units are still not standardised. For instance there is still a difference in what a gill is. It is officially a quarter of a pint but in some places a gill means half a pint.

All Change

At times we only need to do rough calculations with money, but usually it is important to be accurate. This applies not only to giving and checking change, but also to doing sums that involve multiplication and division of money.

LOOKING AT NUMBER

As we saw, multiplication is a quick way of adding up amounts that are the same. If we wanted to know the total cost of 5 items at £17 each, we could add the prices up like this:

£17 + £17 + £17 + £17 + £17 = £85

but a quicker way would be to multiply:

$$\begin{array}{r} £17 \\ \times \quad 5 \\ {\scriptstyle 3} \quad \\ \hline 85 \end{array}$$

(It would be even quicker to use a calculator or, in a shop, a cash till!)

Calculations with money often involve pence as well as pounds. If the price of the item increases to £17.39 we can work out the new total by using exactly the same method. We set it out like this:

$$\begin{array}{r} £17.39 \\ \times \quad 5 \\ {\scriptstyle 3\ 1\ 4} \quad \\ \hline £86.95 \end{array}$$ and multiply by 5 starting with the digit at the right-hand side.

To separate the pounds from the pence a decimal point '.' is used. (There is more about the decimal point in section 10.) You will notice that in the price there are two digits to the right of the decimal point (which are the pence) and two digits to the right of the decimal point in the total.

Here is another example, using larger numbers. A store orders seven video recorders, each costing £344.95, and needs to know the total cost. We can set the sum out like this.

$$\begin{array}{r} £344.95 \\ \times \quad 7 \\ {\scriptstyle 3\ 3\ 6\ 3} \quad \\ \hline £2414.65 \end{array}$$

Always start by multiplying the digit at the right-hand side first, and then work towards the left. Remember to carry forward numbers to the next column to the left when necessary. (Look back at section 5 for a reminder.)

T<small>IP</small>

It often helps to put the point separating the pounds and the pence in the answer space before you start multiplying. Line it up with the point in the top line, in the same way that you line up the digits in columns.

$$
\begin{array}{r}
£26.68 \\
\times \qquad 4 \\
\hline
\end{array}
$$

Now try these
The answers are on page 93; or use a calculator to check the answers.

1 (a) £8.50 × 3
    (b) £26.45 × 4
    (c) £167.92 × 5
    (d) What is the total cost of six bags of cement if each bag costs £17.65?
    (e) If a High Street shop orders seven colour TVs at £274.90 each, what is the total cost?

LOOKING AT NUMBER

Sometimes we have to divide money exactly: perhaps we need to share out pocket money to children, or even share out a pools win!

> Well, the good news is that we've won the pools...

> But the bad news is that we only won £48·20 and the terrible news is that I can't divide that between the 17 of us!

The method explained in section 5 works for pounds and pence as well as whole pounds.

If a restaurant collects £7.35 in tips for one night and it has to be shared between three waiters, we could set it out like this:

$$
\begin{array}{r}
£2.45 \\
3 \,\overline{|\, £7.35}
\end{array}
$$

Each waiter would get £2.45.

Here is another example. If someone gets a Giro cheque for £71.34 every fortnight how much do they get each week? We can work this out by dividing £71.34 by two:

$$
\begin{array}{r}
£35.67 \\
2 \,\overline{|\, £71.34}
\end{array}
$$

This works out at £35.67 a week.

TIP
Again it is a good idea to put the point separating the pounds and pence in the answer space before starting the sum.

$$
\begin{array}{r}
. \\
6 \,\overline{|\, £43.80}
\end{array}
$$

Now try these
The answers are on page 93.

2 (a) £8.24 ÷ 2
    (b) £7.62 ÷ 3
    (c) £23.65 ÷ 5
    (d) If three bags of fish and chips cost £5.85 how much does one bag cost?
    (e) A 5-litre tin of paint costs £12.95 and a 2-litre tin costs £5.80. Which is the better value? (Work out the cost of one litre for each tin.)

LOOKING AT NUMBER

Calculators and cash tills are often used for working out sums with money but, as we saw before, if you don't press the right buttons you don't get the right answers.

> Now then, a small loaf of bread and ½ dozen eggs that's £87·00.

> Blimey! Talk about inflation!

If you wanted to work out £126 ÷ 9, but pressed the × button instead of the ÷ button by mistake you would get the answer £1134 instead of £14!

£126 × 9 = £1134

£126 ÷ 9 = £14

It is a good idea to try to do a rough check in your head. Then you could tell straight away that the answer here must be wrong, because if you divide up a sum of money the parts must be less than the total amount you had to start with. If you start with £126 and share it out 9 ways the answer can't be more than £126, so £1134 must be wrong.

You can get a better idea of what the answer should be by rounding the numbers. If we say that £126 is roughly £130, and 9 is roughly 10, then an approximate answer can be worked out by dividing £130 by 10. Since a quick way of dividing by 10 is to drop the last 0, £130 divided by 10 is £13. The answer should be roughly £13.

If we wanted to multiply £209 by 11, but divided by mistake the answer would be £19 instead of £2299:

£209 ÷ 11 = £19

£209 × 11 = £2299

We can do similar rough checks. When multiplying, we should end up with more than we started with, so £19 must be wrong since it is less than £209. To get a better idea of the answer we can round the numbers to make them easier to work with. We can round the 11 to make it 10 and we can round the £209 to £210. We can multiply by 10 by simply adding a 0 to the number. So £210 × 10 is £2100. The answer should be around £2100.

TIP
When you are working out a rough answer for a multiplication sum it is best to round one number up and one down, but for a division sum either round both numbers up or both numbers down. The most

important thing is to round so that you have numbers that are easy to work with.

Now try these
Work out an approximate answer to these sums, and then use a calculator to work out the exact answer. The exact answers are given on page 93.

3 (a) £171 ÷ 9
(b) £165 × 11
(c) £176 ÷ 11
(d) £155 × 9
(e) The cost of hiring a coach for a works outing is £105. There are 21 people who want to go on the trip. How much will each person need to pay?
(f) In a month, a record shop sells 105 copies of a double album costing £11. What are the total takings for this record in the month?

SIDENUMBER
Computers, calculators, cash tills, automatic check points and cash points – there is a dazzling array of new gadgets to work out problems. But calculators have been around for thousands of years. One of the earliest types, the abacus, is still widely used in the Far East – not just by shopkeepers, but by big businesses as well.

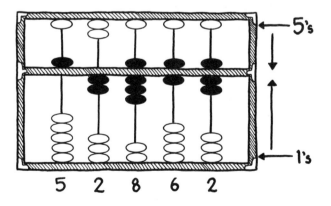

Skilled abacus users can be at least as quick at solving number problems as someone using an electronic calculator!

Computers aren't a new idea either. A Victorian mathematician called Charles Babbage designed an 'analytical engine' which had many of the features of a modern computer. The British government supported Babbage because they saw his machine as a way of working out ranges for artillery. Unfortunately for Babbage, his design relied on mechanical parts and was too difficult to make at the time. After ten years the government decided that it could not give him any more money. It was only with the development of electronics in the 1940s that his ideas became practical. Babbage is now recognised as the 'father of computers'.

If Babbage is the father of computers, then perhaps Ada Lovelace, the daughter of the poet Lord Byron, should be called the 'mother of computers'. Not only did she help Babbage financially but, as a mathematician herself, she worked out how the 'engine' could be 'programmed'. There is now a computer programming language called ADA which was named in her honour.

What's the Point?

So far we have looked at whole numbers, but sometimes we also need to deal with parts of numbers.

You have probably heard of decimals but may be confused about what they are.

LOOKING AT NUMBER

We saw earlier that our number system is based on ten and that increasingly we measure using the metric system, which is also based on ten. We count in tens and multiples of ten:

1	one
10	ten
100	one hundred
1000	one thousand
10000	ten thousand

and so on.

Place value means that a number written in one column is worth ten times what it would be in the column to the right of it. We have seen that we can multiply by ten by simply adding a 0 to the number. This has the effect of moving the digits one column to the left. We have also seen that if we divide a number by ten it has the effect of moving the digits one column to the right.

This is all right for whole numbers, but how does it work with numbers smaller than one? Here we put extra columns at the right-hand side of the number.

In the same way that we can divide a hundred up into ten parts to give us ten tens, and we can split ten into ten ones, so we can split one into ten parts, and we can split each of those parts into ten parts, and so on. So we can write:

1000	one thousand
100	one hundred
10	ten
1	one
0.1	this is one tenth
0.01	this is one hundredth

and so on.

Here the 1 moves one column to the right each time. This is the same as dividing by ten. But we do not stop when we reach the ones column; some extra columns are added on the right-hand side.

This process of splitting into ten parts is called the decimal system. You will notice that there is a '.' which separates the whole numbers from the parts. This is called a decimal point. You see this most often with money, where the dot separates the pounds from the parts of a pound (pennies).

Remember that numbers get their size or value from where they are placed. Just as the 3 in 36 stands for 3 tens, so the 3 in 0.36 stands for 3 ten*ths*. And just as the 2 in 249 stands for 2 hundreds the 2 in 0.02 stands for 2 hundred*ths*.

We would need 10 0.1s to make 1 and we would need 100 0.01s to make 1:

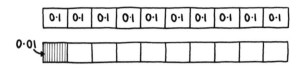

TIP
If in doubt about decimals, think of money: there are 100p in one pound. We can write 10p as a decimal of a pound – £0.10. Similarly:

20p	can be written	as	**£0.20**
50p		as	**£0.50**
75p		as	**£0.75**
1p		as	**£0.01**
5p		as	**£0.05**
746p		as	**£7.46**

Now try these
Write these as decimals of a pound. The answers are on page 93.

1 (a) **60p** (b) **30p** (c) **25p**

(d) **27p** (e) **6p** (f) **999p**

LOOKING AT NUMBER
Of course decimals are not used only for dealing with money. We can use them for measuring length, weight, volume, or anything. We can do sums with decimal numbers just as with whole numbers. We can add, take away, multiply and divide in much the same way.
Look at these additions

Just as it is important to add up whole numbers in columns –

$$
\begin{array}{r}
243 \\
+\ \ 67 \\
\hline
310 \\
\end{array}
$$

the same is true of decimals –

$$
\begin{array}{r}
4.53 \\
+\ 12.67 \\
\hline
17.20 \\
\end{array}
$$

Line up the numbers with the decimal points under one another; the decimal point in the answer goes underneath the other decimal points.

TIP
Think of the decimal point as the barrier between whole numbers and parts of numbers. It may help to leave a large space around the decimal point, and then write the numbers in the correct columns. For example:

$$
\begin{array}{r}
2\ .\ 70 \\
+\ 10\ .\ 05 \\
\hline
12\ .\ 75 \\
\end{array}
$$

Now try these:
Add up the following. The answers are on page 93.

2 (a) £
$$
\begin{array}{r}
6.45 \\
+\ 2.32 \\
\hline
\end{array}
$$

(b) £
$$
\begin{array}{r}
3.29 \\
+\ 7.41 \\
\hline
\end{array}
$$

(c) £
$$
\begin{array}{r}
14.26 \\
+\ 6.92 \\
\hline
\end{array}
$$

(d) 7.10 m.
$$
\begin{array}{r}
+\ 4.91\ \text{m.} \\
\hline
\end{array}
$$

(e) 5.13 kg.
$$
\begin{array}{r}
+\ 8.28\ \text{kg.} \\
\hline
\end{array}
$$

(f) 17.24
$$
\begin{array}{r}
+\ 6.96 \\
\hline
\end{array}
$$

LOOKING AT NUMBER
When taking away decimals it is important to keep them in columns, as with whole numbers:

$$
\begin{array}{r}
545 \\
-\ 265 \\
\hline
280 \\
\end{array}
\qquad
\begin{array}{r}
5.42 \\
-\ 3.65 \\
\hline
1.77 \\
\end{array}
$$

Line up the decimal subtraction with the decimal points underneath one another. To borrow, do this as if the decimal point was not there. Again, the decimal point in the answer lines up with the other decimal points.

Now try these:
The answers are on page 93.

3 (a) £
   4.86
 − 2.64

(b) £
   12.73
 − 4.50

(c) £
   8.72
 − 4.55

(d) 5.89 m.
 − 2.43 m.

(e) 12.65 kg.
 − 6.38 kg.

(f) 10.00
 − 4.56

LOOKING AT NUMBER

With multiplication the process is again similar to multiplying whole numbers; the only problem is deciding where the decimal point goes in the answer. There is a simple rule for this. Count how many digits are to the right of the decimal point in the numbers being multiplied, and then have that many digits to the right of the point in the answer.

      3.9   1 digit to the right
 ×   4   of the decimal point.
     ³
     15.6

      2.15   2 digits to the right
 ×   5   of the decimal point.
      ²
     10.75

Now try these
The answers are on page 94.

4 (a) £3.49
 ×   8

(b) 5.54 m.
 ×   6

(c) 15.03 kg.
 ×   5

LOOKING AT NUMBER

It is possible to do simple division of decimals by hand, but more difficult ones are best done on a calculator.

With a simple division, set out the sum as with whole numbers. The point to watch out for is the decimal point. The decimal point in the answer must be lined up with the decimal point in the number we are dividing.

$$\begin{array}{r} 1.45 \\ 5\overline{)7.25} \end{array}$$

Sometimes, when dividing, we still have a number to carry when we come to the end of the digits in the sum. To get round this, put extra 0s at the right-hand side if necessary. Putting 0s on the end of a number after the decimal point does not alter the number. 7.50000000 is the same as 7.5.

If we want to divide 1.5 by 4 we can start off as normal:

$$\begin{array}{r} 0.3 \\ 4\overline{)1.5} \end{array}$$

but we still have 3 to carry so we add some 0s.

$$\begin{array}{r} 0.375 \\ 4\overline{)1.500} \\ 1.2 \\ \hline 30 \\ 28 \\ \hline 20 \end{array}$$

Now try these
The answers are on page 94.

5 (a) **6.3 ÷ 3**

(b) **3.65 ÷ 5**

(c) **2.6 ÷ 8**

SIDENUMBER

We are used to counting in tens, and probably started doing this because we have ten fingers. The word 'digit' that we use for numbers comes from the Latin word for finger.

Computers don't have fingers, so there is no reason for them to count in tens. The computer is made up of electronic parts, which can be in one of two states either 'on' or 'off'. So for computers it is most convenient to count in twos: this is called the binary system.

Counting in twos is actually no stranger than counting in tens, although it may appear rather odd to us. When we count in tens we split the number up into ones, groups of ten ones (tens), groups of ten tens (10 × 10 or hundreds), groups of ten hundreds (10 × 10 × 10 or thousands), and so on.

When we count in twos we split the number into ones, groups of two ones, then groups of 2 × 2, groups of 2 × 2 × 2, and so on.

When we write a binary number we only need to use 0 and 1. We would write the number 5 in binary as 101, since it is made up of one group of 2 × 2, no groups of 2, and 1 extra one. The number 6 would be 110 – one group of 2 × 2, one 2, and no extra one.

Some of the Parts

Parts of numbers are called fractions. Decimals are just one way of writing and thinking about these. What we have looked at so far (where we split things up into ten parts) are decimal fractions, but there are other ways of dealing with fractions.

LOOKING AT NUMBER

We saw that £0.10 is ten pence, one tenth of a pound. We can write one tenth as 0.10 (or 0.1 which is the same thing) or as $\frac{1}{10}$.

With this way of writing fractions the bottom number says how many parts a whole thing is split up into, and the top number says how many of those parts we have. The parts that we split the thing up into must all be the same size.

Let's look at some common fractions:

$\dfrac{1}{2}$ means that the whole thing is split into 2 parts and you have 1 of them (a half).

$\dfrac{1}{4}$ means that the whole thing is split into 4 parts and you have 1 of them (a quarter)

$\dfrac{3}{4}$ means that the whole thing is split into 4 parts and you have 3 of them (three-quarters).

The whole thing can be anything. When we talk of half a pint of beer the whole thing is a pint of beer. When we talk about three-quarters of an hour the whole thing is an hour. A thing can be split into any number of parts. It is the bottom number that shows how many parts it is split into. The top number shows how many of those parts we have.

Below is a diagram of fractions. The whole thing that we are splitting up is shown as a rectangle, or oblong, but it could be anything.

The more parts you split the whole one into, the smaller each part becomes. This means you need more smaller parts to have the same as one big part. To have the same as one half ($\frac{1}{2}$), you would need two quarters ($\frac{2}{4}$), and you would need four eighths ($\frac{4}{8}$).

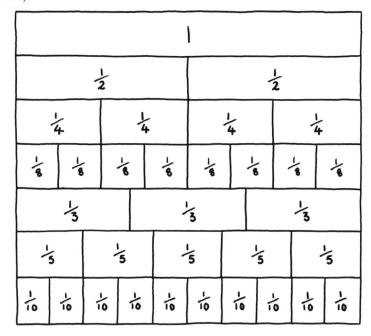

If we think of the whole thing as an hour we can split it into

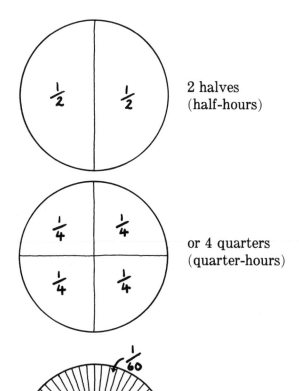

2 halves
(half-hours)

or 4 quarters
(quarter-hours)

or 60 sixtieths
(minutes)

We can see that half an hour is the same as 2 quarter-hours, and is the same as 30 minutes.

$\dfrac{1}{2}$ is the same as $\dfrac{2}{4}$ is the same as $\dfrac{30}{60}$

These fractions may look different, but are actually the same amount. They are called equivalent fractions. We can find an equivalent fraction by multiplying or dividing the top and bottom numbers of a fraction by the same amount:

$\dfrac{2}{4}$ divide both numbers by 2 $\quad\dfrac{1}{2}$ multiply both numbers by 30 $\dfrac{30}{60}$

Dividing both numbers by the same amount is called cancelling down. It is usually more convenient to cancel down fractions where possible. For example it is easier to deal with $\frac{1}{2}$ than $\frac{17}{34}$, although they are actually the same.

Now try these
The answers are on page 94.

1 (a) What do you need to multiply the numbers by to change

$$\dfrac{1}{4} \text{ to } \dfrac{2}{8} ?$$

(b) What do you need to multiply the numbers by to change

$$\dfrac{3}{8} \text{ to } \dfrac{9}{24} ?$$

(c) What do you need to divide the numbers by to change

$$\dfrac{5}{10} \text{ to } \dfrac{1}{2} ?$$

(d) What do you need to divide the numbers by to change

$$\dfrac{30}{66} \text{ to } \dfrac{5}{11} ?$$

LOOKING AT NUMBER
We can add and take away fractions, but this is not as easy as addition and subtraction with whole numbers. We can only add fractions where the parts are the same size.

This cake is split into five equal slices. One slice has a cherry on it, three slices have a candle, and one slice has nothing on it:

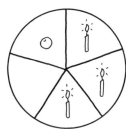

To find how much of the cake has either a cherry or a candle we add up the slices, $1 + 3 = 4$. Since each slice is $\frac{1}{5}$ of the cake, we can write the sum using fractions:

$$\frac{1}{5} + \frac{3}{5} = \frac{4}{5}$$

The bottom number gives the size of the parts (fifths) and the top numbers say how many parts there are. So 1 and 3 give 4 altogether. We can only do this when the parts are the same size.

We can subtract fractions in the same way

$$\frac{3}{4} - \frac{1}{4} = \frac{2}{4}$$

this is the same as

$$\frac{1}{2}$$

since we can divide both the top and bottom number by 2.

Now try these
Work out these sums with fractions. The answers are on page 94.

2 (a) $\dfrac{1}{5} + \dfrac{2}{5}$

(b) $\dfrac{2}{9} + \dfrac{5}{9}$

(c) $\dfrac{1}{8} + \dfrac{3}{8}$

(d) $\dfrac{5}{6} - \dfrac{1}{6}$

LOOKING AT NUMBER
If we want to do sums with fractions, but the parts aren't the same size, we have to use equivalent fractions. For example, to add a half and a quarter, write it out like this:

$$\frac{1}{2} + \frac{1}{4}$$

Because the parts are different sizes we cannot add them up straight away. We have to find an equivalent fraction which has parts of the same size.

We have seen that $\frac{1}{2}$ is the same as $\frac{2}{4}$. So we could write the sum as

$$\frac{2}{4} + \frac{1}{4}$$

and now add these up, to get the answer

$$\frac{3}{4}$$

The use of equivalent fractions for adding and subtracting fractions can get quite complicated, and you will need to look elsewhere, if you want to find out more about it.

TIP
It is often useful to change fractions to decimals, because decimals are usually easier to work with. Here are some common fractions shown as decimals:

$\dfrac{1}{2}$	is the same as	**0.5**
$\dfrac{1}{4}$	is the same as	**0.25**
$\dfrac{3}{4}$	is the same as	**0.75**
$\dfrac{1}{10}$	is the same as	**0.1**
$\dfrac{1}{100}$	is the same as	**0.01**

To change other fractions into decimals, divide the top number by the bottom number. We can see how to do this by looking at $\frac{3}{4}$.

We can convert to a decimal by dividing 3 by 4.

$$\begin{array}{r} 0.75 \\ \hline 4\ \overline{\smash{)}3.00} \end{array}$$

This is the answer shown in the list.

SIDENUMBER

The fractions we have looked at can easily be converted to decimals: $\frac{1}{2}$ is 0.5 and $\frac{1}{4}$ is 0.25 and $\frac{3}{4}$ is 0.75. But there are some fractions which can never be written exactly as decimals.

Let's look at $\frac{1}{3}$: this means 1 divided by 3.

With decimal division we saw that we can add 0s to the right of a number after the decimal point. So we could start this sum

$$\begin{array}{r} 0.333 \\ 3\ \overline{\smash{)}1.000} \\ \underline{9} \\ 10 \\ \underline{9} \\ 10 \end{array}$$

but we could add on 0s for ever, and still not finish it!

The Empire Strikes Back

We have seen how we sometimes use metric measurements and sometimes imperial. Although we often just need to measure approximately – and the best way to do that is to get the look or feel of the units – at other times we need to measure more accurately and to convert one type of measurement into another.

LOOKING AT NUMBER

We have already looked at some approximate ways of converting from one system to the other. The tables in this section give more exact conversions.

This is a table of linear measurement:

Metric	Imperial
1 mm.	0.04 in.
1 cm.	0.39 in.
10 cm.	3.94 in.
50 cm.	19.68 in.
1 m.	39.37 in.
	(3.28 ft.)
10 m.	32.81 ft.
100 m.	109.36 yds.
1 km.	1093.67 yds.
	(0.62 miles)

The conversions are not absolutely exact but are close enough for most purposes.

We can use this table to convert metric into imperial. For instance, to convert 40 cm. into imperial the table shows that 10 cm. is 3.94 inches. As there are 4 lots of 10 cm. in 40 cm. we multiply 3.94 by 4:

$$\begin{array}{r} 3.94 \\ \times \quad 4 \\ \hline 15.76 \end{array}$$

So 40 cm. is the same as 15.76 in.

Now try these
Convert these metric measures into imperial. (A calculator may be useful.) The answers are on page 94.

1 (a) **35 cm.** (b) **4 m.** (c) **1.5 m.**

LOOKING AT NUMBER

To convert imperial into metric we use a similar process:

Imperial	Metric
1 in.	2.54 cm.
1 ft.	30.48 cm.
1 yd.	91.44 cm.
1 mile	1609.34 m.

So to convert 2 ft. 4 in. into metric, take the inches first. One inch is approximately 2.54 cm. and so 4 inches are 4 x 2.54 cm. or 10.16 cm. Then convert the feet. One foot is 30.48 cm. and so 2 feet is 2 x 30.48 cm. or 60.96 cm. We can then add them up: 10.16 cm. + 60.96 cm. = 71.12 cm.

Now try these
Convert these into metric. The answers are on page 94.

2 (a) **10 in.**

(b) **1 ft. 8 in.**

(c) **2 yds. 1 ft. 3 in.**

LOOKING AT NUMBER
We can convert weights in the same way. Remember that there are 16 ounces (oz.) in 1 pound (lb.) and 14 lb. in 1 stone (st.).

Use this table for converting from metric weights to imperial:

Metric	Imperial
1 g.	0.04 oz.
10 g.	0.36 oz.
100 g.	3.53 oz.
1 kg.	2.20 lb.
	(35.28 oz.)

To convert 50 g. to imperial, we can see that 10 g. is the same as 0.36 oz., and since there are 5 lots of 10 g. in 50 g. we can work out the answer by multiplying 0.36 × 5.

0.36 × 5 = 1.8 oz.

So 50 g. is the same as 1.8 oz.

Now try these
Convert these into imperial. The answers are on page 94.

3 (a) **300 g.**

(b) **850 g.**

(c) **4.5 kg.**

LOOKING AT NUMBER
Here is a table for converting imperial weights to metric:

Imperial	Metric
1 oz.	28.35 g.
8 oz.	226.80 g.
1 lb.	453.59 g.
	(0.45 kg.)
7 lb.	3.17 kg.
1 st.	6.35 kg.

(Remember that 1000 grams make 1 kilogram.)

If a man weighs 12 st. 6 lb., we can work out his metric weight by seeing that 1 st. equals 6.35 kg., so 12 st. is 12 × 6.35 = 76.2 kg., and 1 lb. equals 0.45 kg. so 6 × 0.45 = 2.7 kg. If we add these together we find that 12 st. 6 lb. equals 78.9 kg. (76.2 + 2.7 = 78.9).

Now try these
Convert these into metric. The answers are on page 94.

4 (a) **9 oz.**

(b) **2 lb. 8 oz.**

(c) **10 st. 7 lb.**

LOOKING AT NUMBER
The weather forecast usually gives the temperature in degrees Celsius (C), but many of us still think in degrees Fahrenheit (F).

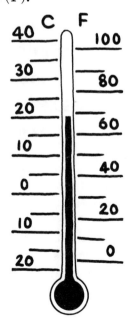

A thermometer measures temperature. We can read off both the Celsius and Fahrenheit scales and convert one to the other. When reading scales like this, go to the nearest line. Here the thermometer shows a temperature of 20°C or 68°F.

Now try these
Using the thermometer, change these to Fahrenheit. The answers are on page 94.

5 (a) **10°C**

(b) **32°C**

(c) **0°C**

and change these to Celsius:

6 (a) **100°F**

(b) **60°F**

(c) **86°F**

LOOKING AT NUMBER

If you want to convert degrees Celsius to degrees Fahrenheit and you haven't got a thermometer you can do it this way:

1 Divide by 5   e.g. **25°C ÷ 5 = 5**

2 Multiply by 9   **5 × 9 = 45**

3 Add on 32   **45 + 32 = 77°F**

If, on the other hand, you want to convert degrees Fahrenheit to degrees Celsius, do this:

1 Take away 32   e.g. **50°F − 32 = 18**

2 Divide by 9   **18 ÷ 9 = 2**

3 Multiply by 5   **2 × 5 = 10°C**

SIDENUMBER
Most everyday thermometers have mercury inside a thin glass tube. Heat makes the mercury expand and rise up the tube. Although mercury is a liquid at most ordinary temperatures, it's actually a metal.

The centigrade scale is divided into 100 parts – 'centi' means one hundredth and 'grade' means step or part. Degrees centigrade are now normally called degrees

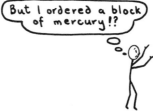

Celsius after the Swiss astronomer, who invented the scale in 1742.

The Fahrenheit system is even older – it dates back to 1714, when Fahrenheit, a German physicist, made his thermometer.

Another very old scale was invented by Réaumur, a French physicist, in 1730. His scale was divided into 80 degrees, and was still used in some countries in very recent times, certainly until 1947.

The Time of Your Life

In section 6 we looked at the 12-hour clock but increasingly time is measured using the 24-hour clock, especially on timetables. There are also larger units of time – days, weeks, months and years – that we need to look at.

LOOKING AT NUMBER

We already know that each day is made up of 24 hours, and that one way of telling the time is to split the day into two lots of 12 hours. With the 24-hour clock we do not split the day in this way. The hours are counted straight through from midnight one night through to midnight the next.

This clock face shows the usual 12 hours

on the outside and, on the inside, the 12 hours that start after midday, or p.m. So the time shown, if it were afternoon, would be either 2:30 p.m. or 14:30.

1 a.m.	2 a.m.	3 a.m.	4 a.m.	5 a.m.	6 a.m.
1:00	2:00	3:00	4:00	5:00	6:00

7 a.m.	8 a.m.	9 a.m.	10 a.m.	11 a.m.	12 a.m.
7:00	8:00	9:00	10:00	11:00	12:00

1 p.m.	2 p.m.	3 p.m.	4 p.m.	5 p.m.	6 p.m.
13:00	14:00	15:00	16:00	17:00	18:00

7 p.m.	8 p.m.	9 p.m.	10 p.m.	11 p.m.	12 p.m.
19:00	20:00	21:00	22:00	23:00	0:00

This table shows how times in the 12-hour clock are written using the 24-hour clock.

When writing times in this clock you write the hours and then the minutes. This table shows that even when there are no minutes you still write 00, to show this. You always use 2 digits to show the minutes, so for 5 minutes past 8 in the evening, you would write 20:05. Sometimes 2 digits are also used to show the hours; so 8 a.m. can be written as 08:00. When times are written down there are various ways of separating the hours from the minutes. Here we have used ':'.

Now try these
Convert these times into the 24-hour clock using either the table or the clock face. (Remember a.m. means before midday and p.m. means after midday, and use two digits to show the number of minutes.) The answers are on page 94.

1   (a) **4 p.m.**   (b) **9 p.m.**

     (c) **8 a.m.**   (d) **5:30 p.m.**

     (e) **10:15 p.m.**   (f) **5 past 10 in the evening.**

Convert these 24-hour clock times into 12-hour clock times using either the table or the clock face. (Remember to put a.m. or p.m. after the time.) The answers are on page 94.

2   (a) **13:00**   (b) **16:30**   (c) **20:15**

     (d) **14:25**   (e) **7:55**   (f) **23:45**

LOOKING AT NUMBER

There are four weeks in a month. There are fifty-two weeks in a year.

So if we divide the number of weeks in a year (52) by the number of weeks in a month (4) we should find out the number of months in a year:

$$\begin{array}{r} 13 \\ 4\overline{\smash{\big)}\,52} \end{array}$$

Thirteen months in a year! Surely there are only twelve?

Both are right. There are 13 lunar months and each one has 4 weeks of 7 days, so each lunar month has 28 days ($4 \times 7 = 28$). But we usually work in calendar months – January, February, March and so on – and most of these have more than 28 days. This table shows how many days there are in each of the 12 calendar months:

January	31
February	28
March	31
April	30
May	31
June	30
July	31
August	31
September	30
October	31
November	30
December	31

TIP

It is not easy to remember how many days there are in each month. This old rhyme can help:

Thirty days hath September

April, June and November

All the rest have thirty-one

Excepting February alone

Which has twenty-eight days clear

And twenty-nine in each leap year

LOOKING AT NUMBER

A year is the time it takes for the earth to go right round the sun.

How many days are there in a year? If you said 365, you would be right, but not completely right. Unfortunately the earth takes about $365\frac{1}{4}$ days to go round the sun and so every fourth year we have to add an extra day: this is a leap year. The extra day is added to February to make 29th February.

TIPS

When is a leap year? It falls every four years just like the Olympic Games. In fact it falls on the same year as the Olympics. So there is an extra day (29th February) every year that the Olympic Games are held. Another way of knowing when a leap year is is to divide the year by four: if it will divide exactly, like 1992, then it is a leap year. But even giving February an extra day does not quite solve the problem of there not being a regular number of days in a year. So an extra adjustment is needed: every year that can be divided by 400 is not a leap year. So the year 2000, even though it is an Olympic year, is not a leap year because it can be divided by 400.

LOOKING AT NUMBER

The days of a month on a calendar are often written like this

JULY 1991

SUN	MON	TUE	WED	THU	FRI	SAT
	1	2	3	4	5	6
7	8	9	10	11	12	13
14	15	16	17	18	19	20
21	22	23	24	25	26	27
28	29	30	31			

To find the date of a particular day, find the column with the name of the day at the top and then look down to see which dates fall on that day. (You will notice that often just the first three letters of the name of the day are used.)

Suppose Mrs Howard left home to go on holiday on the second Thursday in July and stayed away for 10 nights, what date would she get home? By looking down the column headed THU you can see that the second Thursday is the 11th. Then count on ten nights, starting with the 11th. This shows that Mrs Howard's last night away will be the 20th, and so she gets home on the 21st. Looking up the column with the 21st in it you can see that this is a Sunday.

On calendars, the important thing to remember is to look down the columns and across the rows.

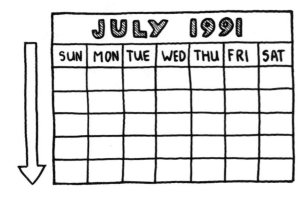

Now try these

Using the calendar, try to answer these. The answers are on page 94.

3 (a) You arrange to meet someone on the first Sunday in July 1991. What will the date be?

(b) You leave home on 12 July. What day of the week is that?

(c) On what day does 3rd August fall?

LOOKING AT NUMBER

Sometimes we want to do sums involving dates. For example we may want to work out in what year someone was born when we know how old they are, or what the date will be in a certain number of years' time. These are just simple additions and subtractions, although the numbers appear to be quite large.

If Robin is 28 in 1991, you can work out the year he was born in by taking his age from the date.

$$
\begin{array}{r}
1991 \\
-28 \\
\hline
1963
\end{array}
$$

If you want to find when he will be 50, that will be 50 years in the future from 1963 and you can work it out by adding 1963 and 50.

$$
\begin{array}{r}
1963 \\
+50 \\
\hline
2013
\end{array}
$$

Now try these
The answers are on page 94.

4 (a) When will someone born in 1991 be 21?
  (b) Kenneth is 50 in 1999. When was he born?
  (c) Audrey will be 45 in 1993. She was 27 when she had a daughter. In what year was her daughter born?

SIDENUMBER
Most of the bigger units of time come from observing patterns in the natural world – day, night, lunar months, seasons, years. But it is not as easy as it seems. For instance in Britain we talk of four seasons, but in many parts of the world there are only two or three, and in some places there is just one season.

Well that's not the case either. When it's 1411 (Muslim), it's 5751(Jewish), or 1991 (Christian), and in Japan it's Heisei 3. It all depends what we take as a starting point. In the west we number years after the traditional birth date of Jesus Christ. In Thailand years are numbered after the Enlightenment of Buddha and so the Christian year 1991 is 2534 in Thailand.

Keep it in Proportion

In this section we will consider percentages, along with the related topics of ratios and proportions. We often see percentages in everyday life.

Ratios are often used when ingredients have to be mixed together in different amounts. This is quite common in cooking, gardening and building, both at home and at work.

LOOKING AT NUMBER

The term 'per cent' comes from the Latin words meaning 'for each hundred'. When we talk about a certain percentage we mean that many for each hundred. It is easiest to think of percentages in terms of money, since there are 100p in £1, but they can be used for many other things as well.

One per cent, normally written as 1%, means 1 for each 100. So 1% of £1 would mean 1p because there are 100p in £1.

1% of £1 is the same as 1p

10% of £1 is the same as 10p

50% of £1 is the same as 50p

Looking at it the other way round –

25p is the same as 25% of £1

80p is the same as 80% of £1

Now try these

Convert these percentages to pence, and pence to percentages. The answers are on page 94.

1 (a) **75% of £1**

(b) **40% of £1**

(c) **13% of £1**

(d) **30p as a percentage of £1**

(e) **65p as a percentage of £1**

(f) **7p as a percentage of £1**

LOOKING AT NUMBER

So far we have looked at percentages of £1 but we often need to work out percentages of numbers bigger than 1. Again it helps to begin by relating examples to money, but the same method can be used with other numbers.

Suppose we wanted to work out 40% of £2:

40% means 40 for each hundred

£2 is 2 hundred pence

So 40% of £2 is 2 × 40p or 80p

Or if we want to find out what 30% of £60 is, we know that £60 is 60 hundred pence and 30% means 30 for each hundred, so 30% of £60 will be 60 x 30p = 1800p or £18.00.

Tip

VAT is charged at 15%. A quick way to find out how much VAT has to be added to a bill is to think of 15% as 10% + 5%. It is fairly simple to work out 10% since 10 is an easy number to work with. We can work out what 5% is by halving the answer we got for 10%, since 5% is half of 10%.

So if the bill is £12 without VAT, we can work out the VAT by finding

10% of £12 12 × 10p = 120p or £1.20

5% of £12 is half of £1.20 so it is 60p

We then add the two amounts together

£1.20 + £0.60 = £1.80

Now try these
The answers are on page 94.

2 (a) **20% of £50**   (b) **70% of £5**

    (c) **5% of 200**

    (d) **What VAT (at 15%) would have to be paid on a bill of £20?**

    (e) **A meal costs £3.70 and a service charge of 10% is added on top. How much is the service charge?**

    (f) **What would be the total cost of the meal in question (e)?**

LOOKING AT NUMBER
If we want to do percentage calculations which involve awkward numbers it is best to use a calculator. Calculators have a special percentage button to make this easier. To work out a sum such as 48% of 250, do this by multiplying the number by the percentage.

We can do this by pressing the buttons in this order

2 5 0
the number

×
multiplied by

4 8 %
the percentage

It should then show the answer 120.

Now try these
Work out these percentages using a calculator. The answers are on page 94.

3 (a) **34% of 7650 kg.**

    (b) **85% of 3060**

    (c) **18% of 4050**

LOOKING AT NUMBER
Sometimes we need to work out what percentage one number is of another, particularly when we want to work out percentage increases or decreases. It is normally easier to use a calculator for this. For example, if a weekly wage is £150, and it increases by £12, what percentage increase is that?

Work the percentage increase out by multiplying the increase by 100 and dividing by the original amount. The buttons to press on the calculator are:

1 2
the increase

multiplied by 100

÷
divided by

1 5 0
the original amount

This gives the answer 8%.

Now try these

Work these out using a calculator. The answers are on page 94.

4 (a) If a television originally cost £250 and goes up in price by £15, what is the percentage increase?

(b) If a football team has an average attendance of 15 000 in one season and it decreases by 3000 in the next season, what is the percentage decrease?

(c) What is the percentage increase if your weight goes up from 75 kg. to 78 kg.?

LOOKING AT NUMBER

Percentages, like decimals and fractions, are ways of looking at parts of numbers.

This table shows percentages with the decimals and fractions which are the same. For instance, 5% is the same as the decimal 0.05 and the fraction $\frac{1}{20}$.

Percentages	Decimals	Fractions		
1%	0.01	$\frac{1}{100}$		
5%	0.05	$\frac{5}{100}$	or	$\frac{1}{20}$
10%	0.1	$\frac{10}{100}$	or	$\frac{1}{10}$
25%	0.25	$\frac{25}{100}$	or	$\frac{1}{4}$
50%	0.5	$\frac{50}{100}$	or	$\frac{1}{2}$
75%	0.75	$\frac{75}{100}$	or	$\frac{3}{4}$
100%	1	1		

Now try these

Fill in the gaps in the following table. The completed table is on page 94.

5

Percentages	Decimals	Fractions		
3%	0.03			
	0.2	$\frac{20}{100}$ or	$\frac{2}{10}$ or	$\frac{1}{5}$
	0.37	$\frac{37}{100}$		
85%		$\frac{85}{100}$ or	$\frac{17}{20}$	
	0.6	$\frac{60}{100}$ or	$\frac{6}{10}$ or	$\frac{3}{5}$
90%		$\frac{90}{100}$		

LOOKING AT NUMBER

If we want to mix some mortar for general use we find that for every 1 kg. of cement we use we need 1 kg. of lime and 6 kg. of sand. The proportions or ratio would be 1 of cement to 1 of lime to 6 of sand. This can be written as 1:1:6.

Ratios are useful because once we remember them we can make up any amount, no matter how large or small.

Suppose we want to make 24 kg. of mortar. How much cement, lime and sand will we need? The way to tackle this is to add up the parts in the ratio 1 + 1 + 6 = 8 then divide the amount we want by the total number of parts. 24 kg. ÷ 8 = 3 kg. So each part is 3 kg. So we would need 3 kg. of cement, 3 kg. of lime and 6 × 3 kg. = 18 kg. of sand.

The same method applies to other 'mixes', even when the ratios are very different. Concrete can be made in different proportions, depending on its use. One ratio is 1 part cement to 3 parts sand to 6 parts gravel, or 1:3:6. So if we needed 20 kg. of concrete in total, to find out how much of each ingredient to use we add up the parts, 1 + 3 + 6 = 10. Divide the total amount needed by that total 20 kg. ÷ 10 = 2 kg. Each part is 2 kg. So we need 2 kg. of cement, 3 × 2 kg. = 6 kg. of sand, and 6 × 2 kg. = 12 kg. of gravel.

Now try these
Work out the following using these ratios – for mortar 1:1:6 of cement, lime and sand; for concrete 1:3:6 of cement, sand and gravel.
The answers are on page 95.

6 (a) The total amount of mortar needed is 16 kg. How much cement, lime and sand is needed?
(b) 250 kg. of concrete is needed to make a small ramp. How much cement, sand and gravel is needed for the job?
(c) A garden wall has to be repointed with mortar. The total amount of mortar needed is 40 kg. How much cement, lime and sand is needed for the job?

TIP
For people who work with dangerous chemicals and medicines, it can be a matter of life and death to work out ratios very accurately. But for most jobs, it is not so important to be exact. For example good cooks do not usually bother to work out the ratios of ingredients accurately. Their experience gives them the right 'feel' for the amounts.

LOOKING AT NUMBER
Ratios are not just used with weight, of course. Sometimes they relate to volume and are often just rough guides to the mix. For instance to make a general-purpose compost we can mix soil, peat and sand in the ratio 7:3:2. We can use any convenient container to measure the volumes, depending on how much we need. It could be a barrowful, a bucketful, or even a handful, since the exact proportions are not critical. If we needed about 6 barrowfuls of compost, we can easily work out the amounts of each ingredient.

We add up the parts in the ratio: 7 + 3 + 2. The total is 12.

Then divide the amount we want by the number of parts: $6 \div 12 = \frac{6}{12} = \frac{1}{2}$.

So each part is $\frac{1}{2}$ a barrowful. We would need $3\frac{1}{2}$ barrowfuls of soil, $1\frac{1}{2}$ barrowfuls of peat and 1 barrowful of sand.

Now try these
Work these out using the ratio for compost of 7:3:2 (soil to peat to sand). The answers are on page 95.

7 (a) We need to fill 24 plant pots with compost. If we measure the ingredients in a plant pot, how many pots full would we need of each?
(b) If a market gardener had four tubs of sand, how many tubs of soil and peat would he need to mix with them to give the right mixture? How many tubs of compost would this make?

LOOKING AT NUMBER
So far the ratios we have looked at involve three different ingredients, but ratios can involve any number of ingredients.

Sometimes it is necessary to work out ratios using different measuring units to the ones given. For instance, a canteen worker buys ingredients in metric units, but wants to use a recipe that is printed in imperial units. The original recipe has the following amounts:

2 lb. potatoes $\frac{1}{2}$**lb. onions**

1 lb. tomatoes   **1 lb. carrots**

Taking $\frac{1}{2}$ lb. as the basic part, the ratio of the ingredients is 4:2:1:2 or 4 parts of potatoes to 2 parts of tomatoes to 1 part of onions to 2 parts of carrots. If the canteen worker wants to make up the recipe using 18 kg. of vegetables the amounts of each can be worked out.

Add up the parts in the ratio 4 + 2 + 1 + 2 = 9.

Divide this into the amount required 18 kg ÷ 9 = 2 kg.

Each part is 2 kg. So the recipe needs 8 kg. of potatoes, 4 kg. of tomatoes, 2 kg. of onions and 4 kg. of carrots.

SIDENUMBER

Although we may think of music as emotional and artistic, and maths as cold and logical, there is actually a close relationship between the two. Many mathematicians are interested in music, and many musicians are interested in maths.

All rhythm is simply a way of representing number, not with figures but with sound. Musical scores give the beat, or tempo, in terms of numbers, 4/4 time or 2/4 time or whatever. Music is written in a way where the length of time of a note is expressed as a fraction, but it is written as different symbols.

The link between music and number does not end there. The lengths of strings on a stringed instrument such as a guitar or piano determine the note, and the notes which harmonise with each other have string lengths which are in direct whole-number ratio to each other. If the lengths of the strings are kept in the right ratio one to another, lower or higher sets of notes harmonise.

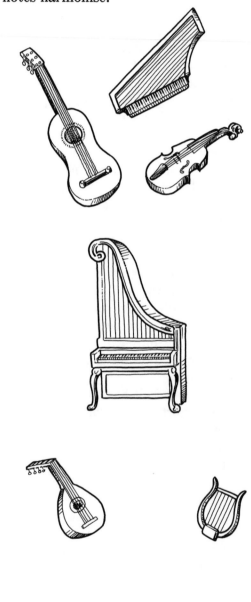

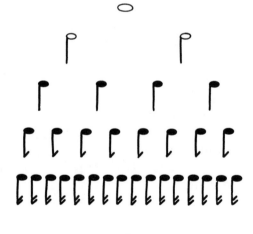

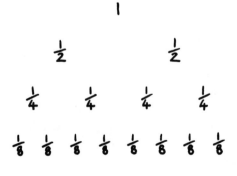

Pies and Bars

It can be hard to imagine numbers. We know that 6 is more than 4 and we know that 80 is more than 50, but it often helps to have a picture of them to see how much bigger they are. We can represent, or picture, 6 and 4 in a number of ways:

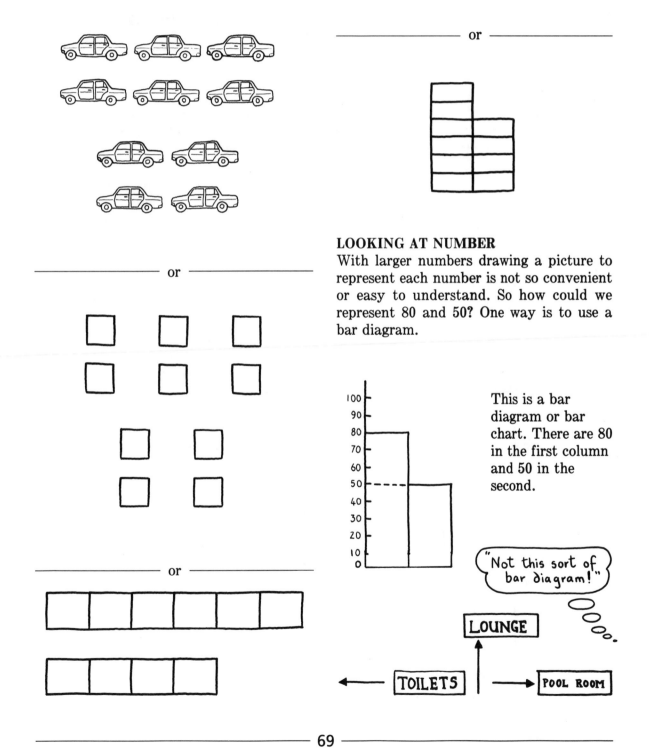

or

LOOKING AT NUMBER

With larger numbers drawing a picture to represent each number is not so convenient or easy to understand. So how could we represent 80 and 50? One way is to use a bar diagram.

or

This is a bar diagram or bar chart. There are 80 in the first column and 50 in the second.

"Not this sort of bar diagram!"

LOUNGE

TOILETS   POOL ROOM

With bar diagrams a column or bar represents each number, and the height of the bar shows the size of the number. To find the size of a column, line up the top with the numbered line which is drawn alongside the bars, and read off the number. This numbered line is called the scale, and the bottom of the scale is called the origin. The origin is usually at zero, and the scale goes up in regular steps.

Sometimes bar diagrams are drawn with the lines going across instead of up and down.

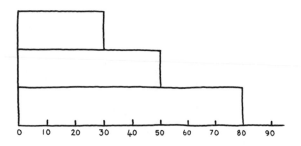

Now try these
Read the amount in each column. The answers are on page 95.

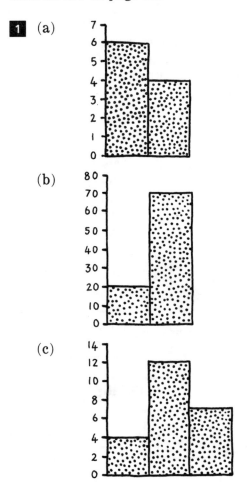

1 (a)

(b)

(c)

LOOKING AT NUMBER

It is quite easy to make a bar diagram. Let's show the following set of numbers in a more interesting way: sales of TVs from a shop were 10 in October, 14 in November and 26 in December. We could show this as

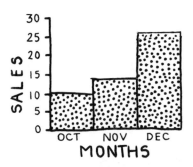

We can see at a glance that there was a massive increase in TV sales in December.

TIP
It is easier to use squared paper to draw a bar diagram. You can buy squared paper or graph paper, which is specially made for drawing graphs and diagrams. If you haven't any squared paper or graph paper you can still draw bar diagrams, but make sure you measure the scale of numbers up the side carefully.

Now try these
Try drawing bar diagrams for these. The answers are on page 95. The size of the total diagram doesn't matter – you can make it small or large. What does matter is that each of the columns can be correctly read off on the scale.

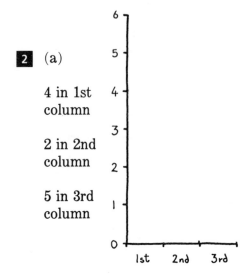

2 (a)

4 in 1st column

2 in 2nd column

5 in 3rd column

(b)

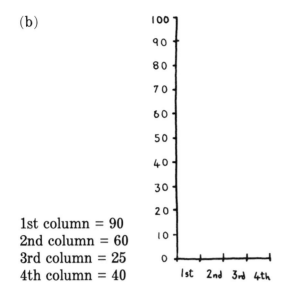

1st column = 90
2nd column = 60
3rd column = 25
4th column = 40

LOOKING AT NUMBER

Another way of representing numbers in a visual way is the pie chart. It's called a pie chart because it looks a bit like a pie cut into slices.

When we looked at fractions we saw that we could show them like this

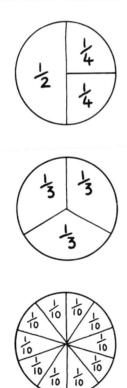

And we can use these 'pies' to help show numbers. If we want to show that in a group of four people three were women and one was a man, we can draw a circle

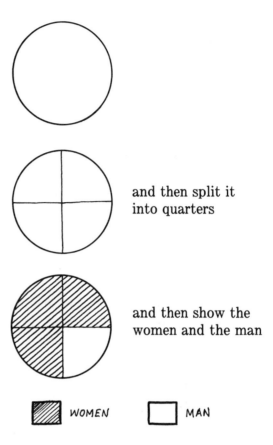

and then split it into quarters

and then show the women and the man

WOMEN   MAN

If there are 12 vehicles and 3 are vans, 3 are lorries and 6 are cars, we can show it on a pie chart in this way:

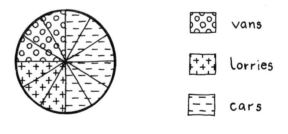

vans

lorries

cars

Any amount of things or items can make up each slice of the pie. And there can be any number of slices in each pie.

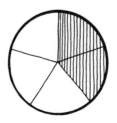

Here 2 slices out of a total of 5 have been shaded in. This shows us that two-fifths of the pie is shaded, and is called a proportion. Each slice could represent one thing, or 2, or 3, or, well, any number. But if each slice represents, say, 5 people, then the shaded area represents $2 \times 5 = 10$. And the unshaded area? Yes $3 \times 5 = 15$.

Now try these
Try to read these pie charts. The answers are on page 95.

 (a)

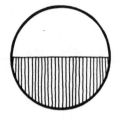

The total pie represents 6 cars. The shaded part represents new cars. How many new cars are there?

(b)

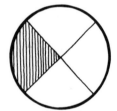

If the whole chart represents 4 million people what does the shaded area represent?

SIDENUMBER
Florence Nightingale is remembered as the 'Lady with the Lamp' – the nurse who kindly visited the sick and dying soldiers during the Crimean War. But that is only a small part of what she did. In fact, she was a very strong woman. She not only formed a professional nursing service, against all the odds, but battled against abuse and corruption in the British army of her day. Medical care throughout the world is still influenced by her actions.

What is not well known is that she helped popularise the idea of representing numbers in the form of pie charts. In order to get money and help to the soldiers she sent constant letters and tables of figures back to the government in London from the war front in the Crimea. She soon realised that these men could not understand all the numbers and the messages they contained. To help them get the point she began to represent, or picture, the amounts in pie charts. She got the help!

Vital Statistics

There is a famous saying 'There are three kinds of lies: lies, damned lies and statistics'. Statistics is the term used for presenting and describing information in the form of numbers, and although statistics are rarely actually lies, it is often possible to present the numbers in a form that suits a particular point of view. Statistics is a vast subject in itself; this section just touches on one or two ideas. It may help you feel more at ease when 'statistics' are presented at work or in the media.

Any information presented in the form of numbers is statistics. If you collected figures on how much you spent each week, these figures would be statistics.

LOOKING AT NUMBER
Rather than listing all of a set of numbers, it is usually easier to try to find just one number which gives an overall feeling of what the numbers are. In everyday speech I might say 'In our house we drink, on average, 2 pints of milk a day' meaning that my family normally drinks 2 pints of milk. Or you might say, 'On average I spend £5 per week on travel', meaning that you may spend more some weeks and less in others, but that it middles out, or averages, at £5 per week.

We can often just look at a group of numbers and decide on one number that can be used to describe them all, but sometimes we need to be more precise. One way to do this is to work out the average, or 'mean', as it is sometimes called.

Here we see the daily amounts a woman spent on travel, rounded up and down to the nearest pound:

MON.	**£5**
TUE.	**£5**
WED.	**£5**
THU.	**£5**
FRI.	**£5**
SAT.	**£10**
SUN.	**£7**

Looking at these figures we can see that most days she spent £5, but on two days she spent more. We could say that on average she spent a bit over £5 a day, but we could be more accurate. To work out the average daily amount spent on travel, add up the daily totals (£42) and divide by the number of days (7). This gives the average spent per day: £6 (42 ÷ 7 = 6).

Whenever an average is needed, simply

add up all the numbers and divide that total by how many numbers there are altogether.

To get the average of 4 6 7 8 5, add them up (30), and divide by how many there are (5) : 30 ÷ 5 = 6. So the average of these numbers is 6.

The same process works no matter how big or small the numbers are, or how many numbers there are. For instance, if three people win the following amounts on the football pools

£7000   and   £3500   and   £1500

the average win is the total (£12 000) divided by the number of winners (3), which is £4000.

Averages are often useful when comparing two sets of numbers. One way to do this is to total each set and then compare the totals, but this can sometimes give a false picture.

If one cricketer has 8 innings, and scores 20, 25, 15, 30, 10, 50, 20 and 30; and another has 6 innings and scores 30, 20, 25, 35, 40 and 30; how can we decide who is the best batsman?

We could total the numbers and find that the first batsman has scored 200 runs, whilst the second one has only scored 180. If we look at the average scores, however, we see that the first batsman has an average of 25 (200 ÷ 8), but the second has an average of 30 (180 ÷ 6).

Now try these
Find the average (mean) of the following numbers. Remember to add up the totals and then divide that total by the number of cases. The answers are on page 95.

1 (a) **2, 6, 1**

    (b) **10, 20, 40, 30**

    (c) **8, 5 , 6, 3, 2, 6**

    (d) **105, 155, 240, 100**

LOOKING AT NUMBER
One area where statistics is widely used is in surveys or sampling. Sampling is a way of finding something out about a whole group of things by only looking at a few of them.

Sampling is very important in industry, particularly in what is called quality control. It is important that the products a factory makes meet certain standards, yet it can be very expensive to go beyond those standards.

For example in a factory that is bottling lemonade it is important to make sure that there is enough lemonade in each bottle, but if there is too much in each bottle the factory could lose money. It would not be sensible to check how much is in every bottle, so a batch of bottles is taken regularly to be measured. The average contents of these bottles is then used to see if the bottling machine is underfilling or overfilling the bottles.

Sometimes we want to find the mean of quite large numbers, but there are ways of making the working easier. For example, if the lemonade factory tested ten of its bottles and found that they contained the following amounts in centilitres

102   105   100   103   104   106
100   102   102   and   106

they could find the mean by adding up all the numbers and dividing by 10.

But there is an easier way. When working out an average we can subtract the same amount from each number, work out the average and then add that amount back on.

So with the lemonade bottles we could subtract 100 from each number, to give

2   5   0   3   4   6
0   2   2   and   6

These numbers add up to 30. We can then divide by 10 to give an average of 3, and then add on 100 to find that the average of the original numbers was 103. So the average content of lemonade in the 10 bottles is 103 cl.

Another way of making the working out

simpler is to divide all of the original numbers by the same amount, work out the average of the new numbers, and then multiply by whatever we divided by.

For example, these are the daily takings in a supermarket:

£42 000   £47 000   £43 000

£56 000   £64 000

and   £72 000

We could add up all the numbers and divide by 6 to get the average, but we can make the sum easier by dividing each figure by 1000 first to give

42   47   43   56   64   and   72

If we add these up we get 324. We can divide 324 by 6 to get an answer of 54. We then multiply this by 1000 to find that the average of the original numbers is £54 000.

Now try these
Work out the means of these numbers by making the numbers simpler first. The answers are on page 95.

2 (a) **1005   1007   1008**
1004   1001

(b) **1200   1500   2700**
1300   1800   500

LOOKING AT NUMBER
The mean is only one way of finding one number to represent a whole set of numbers. In statistics there are two other main ways:

Mode   – The number that occurs most often.

Median   – The middle number of a set of numbers which are written in order. So there are as many numbers larger than the median as there are smaller.

For instance, average weekly earnings can be very misleading:

We might think that this means that most people earn £400 a week, but this could well not be the case. As an example, look at the following figures:

£150   £200   £200

£200   £250   £250

£350   £400   £1600

These are the weekly wages of nine people rounded to the nearest £50. We could work out the mean by adding up the numbers and dividing by the number of people £3600 ÷ 9 = £400, but this doesn't really represent the numbers as a whole. It would be more meaningful to give the mode (£200) since that is the most common wage, or the median (£250) since that is what the 'man in the middle' earns.

There will only be a middle figure for the median if there is an odd number of figures. If there is an even number of figures we work out the median by taking the middle two figures, adding them together and dividing by two. When one or two of the numbers are a lot larger or smaller than the rest, the mode or the median usually give a better 'feel' to the figures as a whole.

TIP
Whenever statistics are presented you need to be aware that the person presenting them may be trying to make things seem different to the way you see them.

Now try these

Work out the mean, median and mode for these figures. The answers are on page 95.

3 (a) **6   3   8   3   700**

   (b) **500   400   700   1600**

       1400   300   700

   (c) **103   108   107   104   103**

       107   103   105

SIDENUMBER

We have only looked at a very small part of statistics. One branch of statistics that affects our lives, often without us realising, is probability. Many things in the world appear to be chance events, and probability is the mathematical way of describing chance events. Most people associate probability with gambling, and certainly as far as gamblers are concerned probability is important, but it is far more important than this.

It is not possible to build a perfect machine. At some time any machine will go wrong. A knowledge of the probability of failure of the parts helps the designer to reduce the risk of breakdown. In aircraft many of the parts are duplicated so that if one fails the other will back it up. If a part has one chance in a million of failing, then the chances of both failing together are one in a million million!

In medicine many things are not certain, but probability can help to show what things are likely to cause a particular disease.

With car insurance, probability can show *how* likely it is that a certain type of person will have an accident, and payments can be adjusted accordingly. This is why it is a lot more expensive for a young, inexperienced person to insure a car.

Many things in life are uncertain; probability is just a way of saying *how* uncertain.

Show Me the Way to Go Home

Nearly everyone has to read maps and plans at some time – everything from local street maps to road maps and plans of buildings.

LOOKING AT NUMBER

Maps and plans are like aerial photographs. They give us a bird's-eye view of what is on the ground. Ordinary pictures show us a side view, so it is not always easy to see where things are, or how far apart. Pictures and photographs are not much use for planning a journey.

Because maps and plans are smaller than the real-life things they are showing we need a way of shrinking things down. We do this by using a scale. A scale means that for a particular length in real life we use a smaller length on the map or plan. For example if something is 1 metre long in real life we may draw it as 1 cm. long on a plan so that it fits on to the paper. This is a scale of 1 cm. to 1 m.

This is a plan of a garden. Each centimetre on the plan represents one metre in the real garden. The scale is 1 cm. to 1 metre.

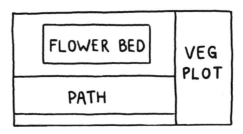

The garden is 6 m. long and 3 m. wide.

Now try these
Measure the following with a ruler and work out how big they are in the real garden. The answers are on page 95.

1 (a) How long is the path?
    (b) What is the area of the flower bed?
    (c) What is the length and width of the vegetable plot?

LOOKING AT NUMBER

When plans are drawn it is important to use an appropriate scale. This will depend both on the size of the piece of paper to be used and the dimensions of what is being drawn. To draw a plan follow these steps:

1   Measure the outside dimensions, the length and width of what you want to draw.
2   Measure the piece of paper.
3   Decide on a scale. Choose a scale that is convenient to use, such as 1 cm. to 1 m. – it would not be very easy to

use a scale of 1 cm. to 1.3 m. Make sure you choose a scale that allows the plan to fit comfortably on the paper.

4   Draw the outside dimensions, to scale.
5   Measure the size and position of the things to be drawn on the plan.
6   Draw them to scale on the plan.

TIP
To draw a plan it is easier to use graph paper, or any paper marked in squares.

Now try this
Draw a plan of one of the rooms in your home. Remember that you are drawing a bird's-eye view. As an example here is a plan of a living room.

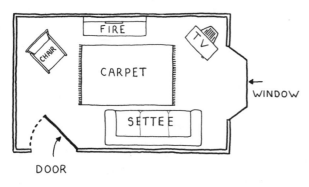

LOOKING AT NUMBER
Maps are drawn in a similar way to plans. Of course they cover a much bigger area, so the scale needs to be very different. The scale may be 1 inch to 1 mile or 1 cm. to 10 km., or even bigger. When using a map it is important to check the key: this shows what the symbols on the map stand for. Here are some symbols commonly used on maps.

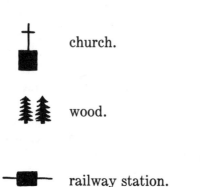

If we want to find a particular street in a town or city we may need to look up the reference to it in the index. If it is in a book of maps or street plans the reference gives the page on which to find the street and it also gives a reference number. This reference number shows where on the page the street can be found.

The map is often divided into squares, with letters and numbers for the rows and columns.

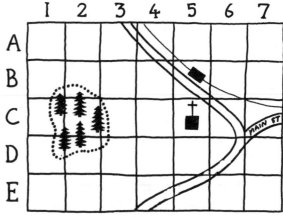

Main Street has a reference of 7C. To find it, run your finger down the column which has a 7 at the top until it crosses the row with a C at the side. This shows which square to look in.

Now try these
Find what is at each of these references. You may need to check the key to find out what the symbols stand for. The answers are on page 95.

2   (a) **5B**   (b) **2C**   (c) **5C**

LOOKING AT NUMBER
With maps and plans we may need to use angles. An angle can be thought of as how far you have to turn from facing in one direction to facing in another. Angles are normally measured in degrees.

Stand facing a wall in your room. Turn right round until you face the wall again. You have turned one complete revolution, or 360 degrees. This is written as 360°. It is made up of four quarter turns or right angles, and so each right angle is 90° (360° ÷ 4 = 90).

The directions, or points on the compass are

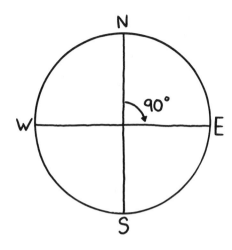

If you stand in the middle facing north and turn to your right until you are facing east you have turned through one right angle, or 90°. If you continue turning until you face south you have turned through two right angles altogether, or 180° (2 × 90° = 180°).

Now try these
Look at these compasses. The arrows show the change in direction. Work out, in both right angles and degrees, how much the direction has changed. The answers are on page 95.

 (a)

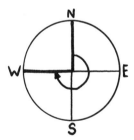

(b)

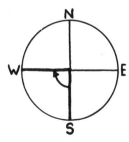

(c)

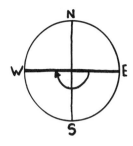

(d)

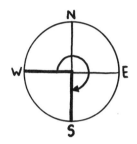

LOOKING AT NUMBER
We talk about the angle between two lines. On a plan the angle between the lines which show two walls of a room might well be a right angle. This means that if we stand in the corner of the room looking along the line of one wall, and then turn to look along the line of the other wall we have turned through one right angle.

SIDENUMBER
Ten is the basis of our number system and metrication. The binary system used in computers counts in twos. But there are also other systems of counting, or rather organising numbers into sets or bases.

No one knows for sure, but the earliest systems of organising and grouping together may come from the Middle East or Ancient China, five or six thousand years ago. Those very early systems were not based on the number ten but on sixty. This may not seem a very obvious number to choose, but it is actually very useful. Sixty can be divided by 1, 2, 3, 4, 5, 6 and 10. There is no smaller number which can be divided by all of these numbers. It can also be divided by 12, 15 and 30. Twelve itself is also rather special because it is the smallest number that can be divided by 1, 2, 3 and 4.

Twelve and sixty are still important in the way we count today. There are 12 inches in a foot, and, before decimal money was introduced, there were 12 pennies in a shilling. We have 12 months in a year and a day is split into two 12-hour periods. Each hour is split into 60 minutes and each minute into 60 seconds.

The compass and all angles and directions are based on 360 degrees, which is a multiple of 60 ($6 \times 60 = 360$). When angles are measured very accurately the degrees are split into minutes and seconds. There are 60 minutes in a degree, and 60 seconds in a minute.

Put Me Off at Crewe

Oh Mr Porter what can I do?
I wanted to get off at Birmingham
But they put me off at Crewe.

The old music hall song has a point. Timetables can be very difficult to read. This is partly because they use the 24-hour clock, partly because of the amount of information they contain, and sometimes because of the symbols used. Timetables are just one type of table. Tables are a way of presenting information in a logical manner.

	JAN	FEB	MAR	APR	MAY	JUN	JUL	AUG	SEP	OCT	NOV	DEC
COSTA BOMB	136	112	112	124	128	152	196	196	168	124	112	152
COSTA DEL LAGER	126	102	102	114	118	142	186	186	158	114	102	142
COSTA LOTTA	144	120	120	132	136	160	204	204	176	132	120	160

LOOKING AT NUMBER

Above is a table showing the cost of a week's holiday at three resorts during the different months of the year.

To find out the cost of your holiday, look across the row of the resort you want and down the column of the month to find where they meet. So a week's holiday in Costa Bomb in August costs £196, but in November it is only £112.

Now try these

Using the table, find the cost for a week's holiday for the following. The answers are on page 95.

1 (a) **Costa Lotta in July**

(b) **Costa Bomb in December**

(c) **Costa Del Lager in June**

LOOKING AT NUMBER

Some tables look very complicated and difficult to read but they are basically the same as the one shown here. To find the information you want look across the right row and down the right column. Where they cross is the information you need.

Below is part of a table from a mail-order catalogue showing the cost of extra insurance cover on electrical goods.

To find the weekly cost of extra insurance cover for a washer/dryer spread over 38 weeks, look down the left-hand column until you find the row 'Washer/Dryers'. Then look along the top until you find 38 weeks. Trace your finger down the '38 weeks' column until it meets the 'Washer/Dryers' row. In the square where they met is the weekly payment – £2.00.

TIP

With tables like this one, where many of the categories in the left-hand column are similar, you need to read them all very carefully to make sure you get the right one.

Now try these

Find the weekly cost of extra cover for the following machines. (Be careful – some are very similar.) The answers are on page 95.

2 (a) A Goblin wet/dry vacuum cleaner for 20 weeks

(b) A Hotpoint twin tub for 38 weeks

(c) A Hoover automatic washing machine for 20 weeks

LOOKING AT NUMBER

Many tables, and timetables in particular, use symbols. It is important to check the key to find out what these symbols mean.

Notes
A   Sleeping Car customers only.
b   22 September arr. 1647
c   23 September arr. 1523
e   15 September only.
f   22 September only.
g   Departs from London Euston.
h   Departs from London Euston. From 5 August depart 2200.
j   Departs from London Euston. From 5 August depart 2325.
k   Change trains at Newcastle.
fo   Fridays only
Ⓡ   Train in which reservation is essential. Reservations are free to ticket holders.
✕   Restaurant service of meals including hot food to customers travelling First Class, also Standard providing accommodation is available.
🛏   Sleepers (also seating accommodation unless otherwise stated).
Times in **bold** type indicate a direct service.
Times in light type indicate a connecting service.

All services shown in this timetable are InterCity unless indicated by ●. InterCity trains offer :
– First Class and Standard accommodation.
– Light food and hot and cold drinks (available for all or part of journey).
– Reservable seats available.

Train Service Information
Further information on Rail Services and Fares can be obtained from the following by telephone:–

London Kings Cross 071–278–2477

The British Railways Board accept no liability for any inaccuracy in the information contained in this timetable – which is subject to alteration, especially during Bank Holiday periods.

CATEGORY		Cover	Total cost	CAT NO	20 weeks	38 weeks	50 weeks
Dry Vacuum Cleaner <	Electrolux Hoover	3 yrs	£16.99	R5140	£0.85		
Dry Vacuum Cleaner <	Goblin. Hitachi	4 yrs	£31.99	R5141	£1.60		
Wet / Dry Vacuum Cleaner <	Vax Goblin	4 yrs	£41.50	RT1.01	£2.10	£1.09	
Automatic Washing Machines (excl. Hotpoint / Hoover / Servis)		3 yrs	£62.99	R8325	£3.15	£1.66	£1.26
Automatic Washing Machines Hotpoint / Hoover / Servis		3 yrs	£45.99	R8326	£2.30	£1.21	
Washer / Dryers		3 yrs	£75.99	R8380	£3.80	£2.00	£1.52
Twin Tubs Hotpoint		4 yrs	£51.99	R8390	£2.60	£1.37	£1.04

This timetable shows trains from London to Edinburgh in summer 1990.

Mondays to Fridays

	Kings Cross depart	Berwick arrive	Edinburgh arrive
	0600	1024	1115
✗	0730	——	1220
✗	0800	1159	1250
✗	0900	1250	1341
✗	1000	——	1432
✗	1030	——	1455
✗	1100	1443	1534
	1130		1616
✗	1200	——	1628
	1300	1643	1738
	1330	1728	1819
✗	1400	1812k	1831
	1430fo	——	1932
	1500	1850	1941
✗	1600	——	2036
✗	1700	2048	2139
✗	1800	2154	2249
✗	1830	2241	2332
⊨ R	2203g	——	0433
R	2215g	——	0525
⊨ A	2345g	——	0600

The times given are in the 24-hour clock. The hours and the minutes are not separated. The first two digits of the time are the hours, and the last two digits are the minutes. The first column gives the times when each train leaves King's Cross station in London; the second column shows when trains arrive at Berwick, and the third column shows when they arrive in Edinburgh. There is one row on the table for each train.

The timetable shows that the 11:00 train from King's Cross arrives in Berwick at 14:43 and Edinburgh at 15:34. The symbol on the left shows that this train has a restaurant service. Where there is a line instead of a time this shows that the train does not stop at that station. For example the 11:30 from King's Cross does not stop at Berwick.

Now try these
The answers are on page 95.

3 (a) What time does the 08:00 train arrive in Edinburgh?
　　(b) What is the first train in the afternoon from Berwick to Edinburgh?
　　(c) What is special about the 14:30 train from King's Cross?

TIP
With complicated tables it often helps to use a ruler or the edge of a piece of paper as a guide for reading across the rows and down the columns. Otherwise it is very easy to jump into the wrong row or column.

LOOKING AT NUMBER
For buses and trains that run frequently the timetable is sometimes shortened, as in the example below: This shows that during the day buses run every 15 minutes. To save space on the timetable only the minutes are shown for the buses between 7:00 and 20:00.

In this example the places are shown down the side of the timetable rather than across the top.

Town Hall	6:00	6:30	7:00		15	30	45	00		20:00	20:30
Church Street	6:10	6:40	7:10	and then at	25	40	55	10	until	20:10	20:40
High Road	6:15	6:45	7:15	these minutes	30	45	00	15		20:15	20:45
Park Street Depot	6:20	6:50	7:20	past the hour	35	50	05	20		20:20	20:50

Now try these
The answers are on page 95.

4 (a) What time is the last bus from the Town Hall?

(b) How long does it take to get from Church Street to the depot?

(c) I have just missed the 15:30 bus at the High Road. What time does the next one leave?

(d) What time does the 12:45 bus from the Town Hall arrive at the depot?

SIDENUMBER

There are different time zones round the world. For instance New York is 5 hours behind the British Isles – so when it is 11 in the morning in London it is 6 a.m. in New York. Singapore is 8 hours ahead of Britain: when it's 10 p.m. in Singapore it is only 2 p.m. in Cardiff or Belfast.

Before railways 'brought the time' each place in the British Isles worked out its own time. Standardised time is quite recent.

The Money Go Round

At one time or another we all need to budget to help us decide whether we can afford to buy something or know whether we have any money to spare. The examples in this section show one way of making the money go round!

LOOKING AT NUMBER

In order to budget effectively the first step is to total all the income, or money that comes in. This varies from person to person, but it is best to list everything that comes in regularly. This includes wages, benefits, or interest earned if you have savings. Don't include things you can't rely on.

It is important to choose a period of time that is convenient. The problem is that regular income can come at different times: usually weekly, fortnightly or monthly, but sometimes only once every six months or once a year, or possibly even longer. Budgeting involves working out how much is received in a given period, say a week or month, in order to get an idea of what you can spend.

TIP
If you are on piece work or receive tips or bonuses it is best to average your income. Section 16 explains how to work out an average.

LOOKING AT NUMBER

In section 13 we saw that although there are 12 months in a year, there are 13 periods of 4 weeks. It is important to realise that if you are paid monthly this is not the same as being paid every 4 weeks. A monthly salary of £780 is not the same as earning £195 a week even though £780 ÷ 4 = £195. We can work out weekly pay from monthly salary by multiplying the monthly salary by 12 and then dividing it by the 52 weeks in the year. It is easier to use a calculator to do this.

$$£780 \times 12 = £9360$$

$$£9360 \div 52 = £180$$

So the real weekly income is £180, not £195.

Now try these
The answers are on page 95.

1 (a) If a family gets £176 a week from one wage earner, £86 a week from another and £15.50 from child benefit what is their weekly income?

(b) A man earns £624 a calendar month. How much is this a week?

(c) A woman has a salary of £936 a calendar month and gets £208 a year in interest from a building society. How much is this a week?

LOOKING AT NUMBER

The second part of budgeting is to work out outgoings, the amount of money that goes out. This will obviously be very different for each person, but there are some common types of problem. Some money is spent daily, for example bus fares. Some may be spent every few days, such as shopping. Some, like rent, may be paid weekly, some may be monthly, and some – such as electricity, gas and telephone – may be every 3 months (13 weeks), or quarterly.

It is important to make a list of all of your outgoings, and if possible allow for unforeseen events, such as repairs.

To budget effectively we need to work out the money which needs to be spent in the same period as we have used for income. So if we have worked out income on a monthly basis we need to work out spending on a monthly basis. Certain items, such as fuel bills or shopping, which are likely to vary, may need to be averaged.

TIP
To make it easier to cope with large bills such as electricity or gas, there are now many ways of paying, for example monthly payments, coin meters or savings stamps, so that you can choose the method most convenient for you.

Now try these
The answers are on page 95.

2 (a) A person spends an average of £45 a quarter on gas, £50 a quarter on electricity and £35 a quarter for the telephone. How much is this a week?

(b) A couple have a community charge (poll tax) bill of £390 each for the year, and yearly water rates of £130 between them. How much is this a week between them?

LOOKING AT NUMBER

The final stage of budgeting is to compare income with outgoings. This shows whether there is enough income to meet the expected outgoings.

One problem that may arise, even when there is enough, is that the bills sometimes arrive before the income. This problem of timing is known as 'cash flow' and affects companies as well as individuals. It is not just a question of having enough income overall, but also of having it when it is needed. If possible, try to keep some money aside to allow for this.

Now try these
(a) Work out your weekly income.
(b) Work out your weekly outgoings.
(c) Compare your income and outgoings.

TIP
If you find your outgoings are more than your income, you should seek advice. There may be benefits that you are entitled to which you are not claiming, or there may be ways of reducing your outgoings.

LOOKING AT NUMBER

Many things, such as cars, motor-bikes and electrical goods, may lose value as they get older. This is called depreciation. When working out whether you can afford something, a car say, you need to allow for this in the calculation. To work out the real weekly cost of running a car you need to include the amount of money that you are losing by depreciation. You can get an idea of this by checking the price of a similar car which is a year older. If you want to buy a second-hand car which costs £2000 and a similar model a year older costs £1500 then the weekly depreciation is the difference in price: £500 divided by 52, or approximately £10. This should be added to the other costs of running a car (such as petrol and insurance) to find the true weekly cost.

You can get an idea of car prices by looking at local papers or car magazines. But remember that it is not just age which determines the value of a car.

LOOKING AT NUMBER

Interest is normally worked out in periods of months or years. It is quoted as a percentage (%). There is more about percentages in section 14. If we borrow some money for a year and the rate of interest is 15%, this means that we need to pay back an extra 15p for each £1 borrowed. If we borrow £200 at 15% interest for a year, the interest will be 15p for each pound, or £30 (200 × 15p = 3000p, or £30.00). So we will have to pay back £230 altogether, the £200 we borrowed and £30 interest. You can work out how much this is a week by dividing by 52.

SIDENUMBER
Most countries which use the decimal system for money, such as Britain, have coins or notes that come in values of

1   2   5 and **10**

(In Britain these are 1p, 2p, 5p, and 10p)

sometimes 10 times these

10   20   50 and **100**

(In Britain these are 10p, 20p, 50p, and £1)

or 100 times these

100   200   500 and **1000**

(In Britain these are £1, £2, £5, and £10 although the £2 is not often used)

or 1000 times these

1000   2000   5000 and **10 000**

(These are £10, £20, £50 and £100. We don't actually have a £100 note at the moment)

Why do we have these rather than, say, a 3p coin or a £7 note?

The reason is that 10 can be divided by

1   (10 ones)
2   (5 twos)
5   (2 fives)
10   (1 ten)

Numbers which will divide into another number exactly are called factors. 1, 2, 5 and 10 are all factors of 10. So we can use a whole number of these coins to make up 10p.

If we had a 3p coin we could not use a whole number of these to make up 10p because 3 is not a factor of 10.

Before 1971 money in Britain was based on 12 and 20. There were 12 pennies in a shilling and 20 shillings in a pound. As far as sharing is concerned, 12 is a better number than 10 because it has more factors. It can be divided by

1   (12 ones)
2   (6 twos)
3   (4 threes)
4   (3 fours)
6   (2 sixes)
12   (1 twelve)

So in the old money coins were based on these factors:

 Penny

 Threepence

 Sixpence

Shilling

Numbers whose only factors are one and the number itself are called prime numbers. For example 7 is a prime number because its only factors are 1 and 7. We have seen that 10 is not a prime number because, apart from 1 and 10, it also has factors of 2 and 5.

The first few prime numbers are 1, 2, 3, 5, 7 and 11. Try to work out the next 4 prime numbers. They are listed on page 95.

The Power of Number

This is it! The final section at last. In the previous sections we have looked at how we use numbers in lots of everyday situations, at work, rest and play. But numbers can be powerful.

LOOKING AT NUMBER

Many people see numbers as powerful in themselves: there are a lot of superstitions about numbers. The number 13 is often considered unlucky, and some people will try to avoid it at all costs. The 13th of a month is thought to be an unlucky date, particularly if it is on a Friday. Some streets don't have a house number 13, some hotels don't have a room number 13, even the substitutes in a football team are numbers 12 and 14.

The number 3, on the other hand, is usually thought to be lucky.

Lucky and unlucky numbers are numbers that usually have some religious significance, but some people have their own lucky and unlucky numbers, often because of something that has happened in the past.

But numbers have no power on their own; they are just a means of counting. The real power of numbers is in how they are used.

Many of the things we take for granted would be very difficult, or impossible, without the power of number. Astronauts couldn't have gone to the moon without the use of numbers, governments couldn't govern without the use of numbers, we wouldn't know how to use electricity without the use of numbers, we couldn't design cars without the use of numbers.

ENDNUMBER

We have looked at numbers as tools for solving problems, but numbers can be fascinating in themselves. We can use numbers to perform tricks, we can use numbers to create beautiful patterns, but most of all we can use numbers for fun. Here we will look at a few things we can do with numbers. Some are nothing more than entertainment, but some are actually very useful.

Number Magic

Think of a three-digit number where the first and last digits are different.

Reverse the digits of the number (if you thought of 981 reversing the digits gives 189).

Subtract the smaller number from the larger number.

Reverse the digits of the answer.

Add this number to the answer. You should end up with 1089.

For example

$$\begin{array}{r} 981 \\ -\ 189 \\ \hline 792 \end{array}$$

(reverse the digits and subtract)

$$\begin{array}{r} 792 \\ +\ 297 \\ \hline 1089 \end{array}$$

(reverse the digits of the answer and add)

This will work with any three-digit number where the first and last digits are different. If, however, you end up with a two-digit number when you do the subtraction you will need to put a 0 in front of it before you reverse it. So if you get an answer of 99 from the subtraction you need to write this as 099 and reverse it to give 990.

Try it for yourself.

Integer Sums

To find the integer sum of a number add the digits of the number together. If this gives an answer greater than 9, add the digits of the answer together, and carry on doing this until you end up with just one digit. This is the integer sum.

To find the integer sum of 8762 add up the digits $8 + 7 + 6 + 2 = 23$ and then add up the digits of the answer $2 + 3 = 5$. So the integer sum of 8762 is 5.

Here is an interesting fact about integer sums. If you multiply two numbers together the integer sum of the answer is the same as the integer sum of the answer you get from multiplying the integer sums of the original numbers. This can be seen more clearly in the following example:

$39 \times 17 = 663$

(The integer sum of 663 *is 6*:

$6 + 6 + 3 = 15; 1 + 5 = 6$)

The integer sums of 39 and 17 are 3 $(3 + 9 = 12; 1 + 2 = 3)$ and 8 $(1 + 7 = 8)$

$3 \times 8 = 24$

The integer sum of 24 *is also 6*:

$(2 + 4 = 6)$

This is useful for checking multiplication sums. It won't tell you if you have got the sum right, but it will tell you if you are wrong.

Pascal's Triangle

Blaise Pascal was a seventeenth-century French mathematician, who among other things invented one of the first adding machines. He is probably best remembered because of the pattern of numbers shown below, which is named after him.

Pascal's triangle can be continued for ever. The numbers in the triangle are

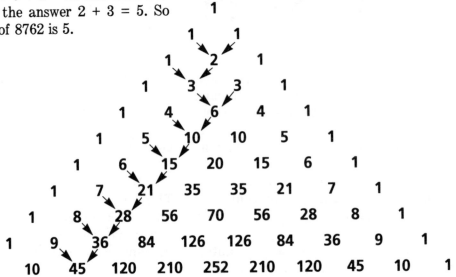

worked out by adding the two numbers above. The triangle contains all sorts of patterns of numbers. Try shading all the even numbers, or all the numbers that divide by 3, and see what happens.

The numbers in the triangle occur in many things in real life. For example, in football pools, if you pick 8 draws from 10, that is 45 goes altogether. The bottom line in the triangle illustrated shows the number of different combinations of selecting a certain number out of ten.

The first number **1** shows the number of combinations of **10** things out of **10**

The second number **10** shows the number of combinations of **9** things out of **10**

The third number **45** shows the number of combinations of **8** things out of **10**

The fourth number **120** shows the number of combinations of **7** things out of **10** and so on.

If we toss two coins we could get two heads, a head and a tail, a tail and a head, or two tails. So there are four different things that could happen. The third line of the triangle tells us how many different combinations there are. There is one combination with two heads, two combinations with one head, and one with no heads. This knowledge of the various possible combinations can be useful in a lot of card games.

Russian multiplication

This is a very different way to do multiplication, but it works.

To multiply two numbers using Russian multiplication we start off by writing the numbers alongside each other, with the smaller number on the left. We then halve the smaller number and double the larger number, and write them on the next line. If the smaller number will not split equally into two then ignore anything that is left over. So we would say that a half of 17 is 8 and forget about the one that is left over. We carry on doing this until we are left with 1 on the left-hand side. We then

cross out all of the lines which have an even number on the left, and add up the numbers that remain on the right, to find the answer. This sounds rather complicated; here is an example.

To multiply 17 × 38 we first write the two numbers:

17	38	
8	76	halve the numbers on the left
4	152	double the numbers on the right
2	304	
1	608	

Then cross out all the lines with an even number on the left-hand side. This leaves just two lines

17	38
1	608

If we add up the remaining numbers in the right-hand column we get the answer 38 + 608 = 646. So 17 × 38 = 646.

Try it yourself, it should always work.

Magic Squares

Magic squares were once a popular pastime. The idea of a magic square is to arrange a series of numbers into a square so that they add up to the same total across, down and diagonally.

Here is a square where the numbers 1 to 16 are arranged in a 4 by 4 square so that each row, column and diagonal adds up to 34.

7	12	1	14
2	13	8	11
16	3	10	5
9	6	15	4

Try to arrange the numbers 1 to 9 in a 3 by 3 square so that each row, column and diagonal adds up to the same amount.

The best of luck! And I hope you have enjoyed this introduction to number.

Answers

SECTION 1

1   (a) 25
     (b) 40
     (c) 362
     (d) 250
     (e) 506
     (f) 900

2   (a) Thirty-one
     (b) Sixty
     (c) Two hundred and ninety-five
     (d) Three hundred and forty
     (e) Four hundred and nine
     (f) Seven hundred

3   (a) 14
     (b) 31
     (c) 77
     (d) 65
     (e) 81
     (f) 84

4   (a) 6 cm.
     (b) 3 cm.
     (c) 5 cm.
     (d) 3 in.
     (e) 2 in.
     (f) 4 in.

5   (a) 20
     (b) 30
     (c) 40

6   (a) 2220 cm.
     (b) 3320 cm.
     (c) 1380 miles

SECTION 2

1   (a) £4
     (b) £50
     (c) £20
     (d) £100

2   (a) 21
     (b) 41
     (c) 20
     (d) 24

3   (a) 46
     (b) 29
     (c) 7
     (d) 15

SECTION 3

1   (a) 15
     (b) 48
     (c) 28
     (d) 36

2   (a) 7 m.^{2}
     (b) £70
     (c) $17\frac{1}{2}$ m.^{2}

SECTION 4

1   (a) 4
     (b) 6
     (c) 8

2   (a) 2
     (b) 6
     (c) 7

3   (a) 10 boxes
     (b) 5 bars

SECTION 5

1   (a) 1306
     (b) 1500
     (c) 8324
     (d) 26299

2   (a) 214
     (b) 378
     (c) 243
     (d) 1679
     (e) 2346

3   (a) 540
     (b) 1248
     (c) 1632
     (d) 67760

4   (a) 124
     (b) 123

(c) 121

(d) 432

5 (a) 145

(b) 119

(c) 120

(d) 32

6 (a) 25

(b) 431

(c) 587

(d) 1247

SECTION 6

1 (a) Seven fifty

(b) Three thirty-five

(c) Nine thirty

(d) Six fifty-five

2 (a) 1:30 one thirty or half past one

(b) 11:00 eleven o'clock

(c) 3:40 three forty or twenty to four

3 (a) 10:25

(b) 6:35

(c) 11:00

(d) 8:15

SECTION 7

1 (a) 1 kg. 200 g.

(b) 3 kg. 630 g.

(c) 4 kg. 300 g.

(d) 9 kg.

2 (a) 10 lb.

(b) 1 lb.

(c) 500 g.

(d) 120 or 125 g.

3 (a) 515 g.

(b) 305 g.

(c) 500 g.

(d) 25 g.

SECTION 8

1 (a) 2 pints

(b) 5 fl. oz.

(c) 2 gallons

2 (a) 30 cl.

(b) 300 cl.

(c) 9 litres

3 (a) 24 m.^{3}

(b) 48 m.^{3}

4 (a) 40 m.p.g.

(b) 30 m.p.g.

(c) 8 gallons

(d) £14.40

SECTION 9

1 (a) £25.50

(b) £105.80

(c) £839.60

(d) £105.90

(e) £1924.30

2 (a) £4.12

(b) £2.54

(c) £4.73

(d) £1.95

(e) 5-litre tin

3 (a) £19

(b) £1815

(c) £16

(d) £1395

(e) £5

(f) £1155

SECTION 10

1 (a) £0.60

(b) £0.30

(c) £0.25

(d) £0.27

(e) £0.06

(f) £9.99

2 (a) £8.77

(b) £10.70

(c) £21.18

(d) 12.01 m.

(e) 13.41 kg.

(f) 24.20

3 (a) £2.22

(b) £8.23

(c) £4.17

(d) 3.46 m.

(e) 6.27 kg.

(f) 5.44

4 (a) £27.92
(b) 33.24 m.
(c) 75.15 kg.

5 (a) 2.1
(b) 0.73
(c) 0.325

SECTION 11

1 (a) 2
(b) 3
(c) 5
(d) 6

2 (a) $\frac{3}{5}$
(b) $\frac{7}{9}$
(c) $\frac{4}{8}$ or $\frac{1}{2}$
(d) $\frac{4}{6}$ or $\frac{2}{3}$

SECTION 12

1 (a) 13.65 in.
(b) 13.12 ft. or 157.48 in.
(c) 4.92 ft. or 59 in.

2 (a) 25.4 cm.
(b) 50.8 cm.
(c) 220.98 cm. or 2.21 m.

3 (a) 10.59 oz.
(b) 30 oz. or 1 lb. 14 oz.
(c) 9.9 lb. or 9 lb. 14 oz.

4 (a) 255.15 g.
(b) 1134 g. or 1.13 kg.
(c) 66.7 kg.

5 (a) 50°F
(b) 90°F
(c) 32°F

6 (a) 38°C
(b) 15°C
(c) 30°C

SECTION 13

1 (a) 16:00
(b) 21:00
(c) 8:00

(d) 17:30
(e) 22:15
(f) 22:05

2 (a) 1 p.m.
(b) 4.30 p.m.
(c) 8.15 p.m.
(d) 2.25 p.m.
(e) 7.55 a.m.
(f) 11.45 p.m.

3 (a) 7th
(b) Friday
(c) Saturday

4 (a) 2012
(b) 1949
(c) 1975

SECTION 14

1 (a) 75p
(b) 40p
(c) 13p
(d) 30%
(e) 65%
(f) 7%

2 (a) £10
(b) £3.50
(c) 10
(d) £3
(e) 37p
(f) £4.07

3 (a) 2601 kg.
(b) 2601
(c) 729

4 (a) 6%
(b) 20%
(c) 4%

5

Percentages	Decimals	Fractions
3%	0.03	$\frac{3}{100}$
20%	0.2	$\frac{20}{100}$ or $\frac{2}{10}$ or $\frac{1}{5}$
37%	0.37	$\frac{37}{100}$
85%	0.85	$\frac{85}{100}$ or $\frac{17}{20}$
60%	0.6	$\frac{60}{100}$ or $\frac{6}{10}$ or $\frac{3}{5}$
90%	0.9	$\frac{90}{100}$ or $\frac{9}{10}$

6 (a) 2 kg. of cement, 2 kg. of lime and 12 kg. of sand
    (b) 25 kg. of cement, 75 kg. of sand and 150 kg. of gravel
    (c) 5 kg. of cement, 5 kg. of lime and 30 kg. of sand

7 (a) 14 pots of soil, 6 pots of peat and 4 pots of sand
    (b) 14 tubs of soil, 6 tubs of peat making 24 tubs altogether

SECTION 15

1 (a) 6,4
    (b) 20,70
    (c) 4,12,7

2 (a)

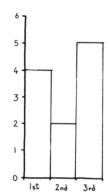

    (b)

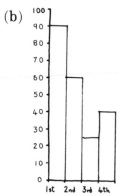

3 (a) 3 new cars
    (b) 1 million people

SECTION 16

1 (a) 3
    (b) 25
    (c) 5
    (d) 150

2 (a) 1005
    (b) 1500

3 (a) mean 144, median 6, mode 3
    (b) mean 800, median 700, mode 700
    (c) mean 105, median 104.5, mode 103

SECTION 17

1 (a) 4.5 m.
    (b) 3 m.^{2}
    (c) 3 m. by 1.5 m.

2 (a) Railway station
    (b) Wood
    (c) Church

3 (a) 3 right angles or 270°
    (b) 1 right angle or 90°
    (c) 2 right angles or 180°
    (d) 3 right angles or 270°

SECTION 18

1 (a) £204
    (b) £152
    (c) £142

2 (a) £2.10
    (b) £1.37
    (c) £2.30

3 (a) 12:50
    (b) 12:50
    (c) Fridays only

4 (a) 20:30
    (b) 10 min.
    (c) 15:45
    (d) 13:05

SECTION 19

1 (a) £277.50
    (b) £144
    (c) £220

2 (a) £10
    (b) £17.50
    The next four prime numbers are 13, 17, 19 and 23

The New  City and Guilds

Certificate in Communication Skills (Wordpower)

If you have enjoyed working through this material you may be interested in gaining a qualification in numeracy. The City and Guilds Certificate in Numeracy (Numberpower) shows the skills you have gained in using and understanding numbers and mathematical concepts. It is available at a large number of centres throughout the country. For further information about the centres that offer the certificate contact: ALBSU, 7th Floor, Kingsbourne House, 229-231 High Holborn, London WC1V 7DA.